The Clean Energy Transition

The Clean Energy Transition

Policies and Politics for a Zero-Carbon World

Daniel J. Fiorino

polity

First published in 2022 by Polity Press

Polity Press
65 Bridge Street
Cambridge CB2 1UR, UK

Polity Press
111 River Street
Hoboken, NJ 07030, USA

ISBN-13: 978-1-5095-4486-8
ISBN-13: 978-1-5095-4487-5 (pb)

A catalogue record for this book is available from the British Library.

Library of Congress Control Number: 2022932838

Typeset in 10.5 on 12.5pt Sabon
by Fakenham Prepress Solutions, Fakenham, Norfolk NR21 8NL
Printed and bound in the UK by CPI Group (UK) Ltd, Croydon

The publisher has used its best endeavours to ensure that the URLs for external websites referred to in this book are correct and active at the time of going to press. However, the publisher has no responsibility for the websites and can make no guarantee that a site will remain live or that the content is or will remain appropriate.

Every effort has been made to trace all copyright holders, but if any have been overlooked the publisher will be pleased to include any necessary credits in any subsequent reprint or edition.

For further information on Polity, visit our website:
politybooks.com

For
Matt and Lauren
Jake and Kelly
Andrew and Cam

Contents

Preface

The future of the planet and thus of humanity is tied inextricably to how we produce, use, and rely on energy. Compelling evidence of the effects of climate change is the most obvious sign of this link. The summer of 2021 brought massive floods in Germany and China, searing heat waves in the northwestern United States, an economically costly early frost in France, raging wildfires in Siberia, and damaging drought in much of the world. The longer it takes to cut energy and other emissions, the worse the impacts of a changing climate will be, and the harder it will become to change the emissions trajectory.

A changing climate is just the starting point for considering the effects of the current global energy system. Air pollution from fossil fuels is a leading threat to human health, a cause of illness and premature deaths, and a drag on prosperity. A fossil-fuel-based global energy system causes untold devastation to ecosystems around the world, from leaking pipelines and oil spills to the abandoned coal mines that dot rural landscapes. Moreover, the energy system as it developed over the last two centuries has concentrated economic and political power in ways that undermine social and political equity and stability.

The world now is engaged in the early stages of a transition to a cleaner energy system. This is by no means the first energy transition in history. There have been multiple such transitions: from a reliance on human and animal muscle and primitive wind, water, and solar technologies, to the unfolding of a coal-enabled industrial revolution, to the emergence of the nuclear industry in the mid-twentieth century.

Yet this latest transition is the most consequential, because so much depends on it. The need for the move to clean energy, defined here as getting carbon out of the global system, is unassailable.

The transition to a decarbonized world is underway, but is not moving fast enough. It needs to be accelerated by committed governments, using well-designed, thoughtful policies. Action by individual governments is not enough. It helps, of course, if China and the United States, together responsible for some 40% of climate-related emissions, join forces with the European Union and other advanced economies in a concerted effort to clean up the global energy system. Even then, most emissions growth between now and mid-century – our decarbonization timetable – will occur not only in China but also elsewhere in southeastern Asia, and eventually in Africa and other parts of the world. It is not only the affluent countries that need to mobilize, but also those seeking affluence. And there is a compelling interest in having the developed countries support clean energy in developing ones.

In the last decade, the challenge has been defined more sharply – and more realistically – as one of *net-zero carbon* or *carbon neutrality*. This is the idea that we need to balance the books: to offset any remaining emissions later in this century with enhanced natural and technological means of removing carbon from emissions and even (after the fact) from the atmosphere. But reaching the elusive goal of carbon neutrality will require a fundamental transformation of the global energy system.

The transition to clean energy, then, has to happen globally, and it will not be easy. Energy is so much a part of modern life that it affects how people around the world work, play, move, invest, and survive. As we think about the climate change crisis and the other harms arising from the current energy system, it is tempting to aim for too much, too fast, without thinking it through carefully. The transition to clean energy is complicated, full of uncertainty, and has many moving parts. As with any complex endeavor, it helps to understand the obstacles as well as the opportunities.

A quote attributed to the organizational theorist, writer, and management professor Robert Anthony is apt: "Moving fast is not the same as going somewhere." The current energy transition is so complex, and so much depends on it, that it has to be informed, intentional, and well designed. If our goal is getting carbon out of the global energy system sometime later this century – the goal much of the world has committed to – then there are limited opportunities for do-overs. It is tempting to want to eliminate fossil fuels by 2030, ban all oil and diesel cars by 2032, shut down all nuclear power plants, or make hydrogen the fuel for all international transport within a few decades. But those things are not going to happen, nor should they if a clean energy transition is going to be fair, durable, and successful in achieving its purpose.

The purpose of this book is to inform readers on the challenges and opportunities of a clean energy transition, present the case for a clean energy world, lay out and assess options for getting there, and offer advice on policies for accelerating the process. Many choices are left for the reader to consider: Does nuclear power have a role? How fast should we move? Should we invest in carbon removal technologies or just in renewables? Is electric mobility the economic wave of the future? What is the role of energy efficiency in the transition? How do we balance an electricity grid nearly totally dependent on variable renewables like wind and solar, and supply renewable power to all users?

My goal is to present the key issues and concepts, assess the paths to decarbonizing, lay out the leading policy options, offer perspectives on the available technologies, and shed light on the economic, environmental, health, social, and political issues associated with the clean energy transition.

From an economic and technological perspective, a transition to a far cleaner energy system sometime later this century is entirely feasible. The challenge is the politics. Albert Einstein reputedly once said that "Politics is more complicated than physics." Although political scientists, including myself, might challenge that statement on a personal level, its relevance to clean energy is incontestable. The economic and technology aspects of the clean energy transition are difficult, but the politics are indeed complicated.

Like any policy field, that of energy involves terms and concepts that will be new to most readers. To some degree, it means learning a new policy language. To help the reader, the book includes a glossary of several key energy terms as a guide to this new language. When a term is first used, it is marked in bold in the text. An emboldened term is thus a cue to refer to the glossary for a brief description.

Many people have helped in this project. I received valuable research assistance as well as sound advice in the early stages from Carley Weted, Sabina Blanco Vecchi, and José Sáenz Crespo. They not only provided background research but also offered advice on priority issues and the overall approach. Michele Aquino provided later research support and handled the permissions for materials used in the book. Michael Kraft offered valuable, detailed advice on the manuscript, for which I am very grateful. I also want to thank Tony Rosenbaum and the three reviewers for Polity Press for their comments and suggestions.

I also owe thanks to my undergraduate students on the 'Energy Politics and Policy' course at American University for helping me think through the case for, and the makeup of, the clean energy transition.

At Polity, thanks go to Louise Knight for helping to develop the idea for the book and its approach, and to Inès Boxman for her advice and support throughout the project. Tim Clark provided expert, skillful copy editing, for which I am very grateful, and Rachel Moore guided the book through the production process.

This book is dedicated to Matt and Lauren, Jake and Kelly, and Andrew and Cam. They will see the effects of the clean energy transition as well as the harm should the transition move too slowly. Thanks finally, and as always, to Beth Ann Rabinovich for creating a positive environment for writing about these and so many other issues.

List of Figures, Tables, and Boxes

Figures

Tables

Boxes

1

The Energy Landscape

In September 2020, Ursula von der Leyen, President of the European Commission, announced the European Green Deal, committing the European Union to reducing carbon emissions by "at least 55%" by 2030. By 2050, Europe plans to be the world's first "climate-neutral continent" in the fight against the existential threat of climate change. Failing to act, the announcement stated, will not only exacerbate climate change, it could lead to over 400,000 deaths from air pollution and 90,000 from heat waves annually. "The longer we wait," the EC warned, "the harder it becomes to reach low-temperature targets and the more expensive the necessary efforts will become."[1]

That same month, President Xi Jinping of China announced that the country with the largest climate-related emissions in the world aimed to be "carbon neutral" within the next forty years.[2] Over the next decade, China would begin reducing emissions, which would begin to fall "steeply" after 2035, with **carbon neutrality** achieved by 2060 (when remaining emissions will be balanced out by **carbon removal**). This is ambitious for a country that in 2020 was generating two-thirds of its electricity from coal plants and had another 200 such plants planned or under construction. According to one projection, however, even in 2060 fossil fuels (coal, natural gas, and oil) would still make up 16% of the energy system, which would be rendered carbon neutral by **carbon capture and storage** (CCS) technology and natural sinks like forests.[3] Many experts thought this was doable, but that it would require a doubling of electricity generation with **renewables** and a major expansion in solar, wind, and nuclear power. One model foresaw an electricity mix for China of

28% nuclear, 21% wind, 17% solar, 14% hydropower, 8% biomass, and the rest from coal and gas using CCS.

Europe and China are not alone in this quest to get carbon out of their economies. According to the World Resources Institute (WRI), by the close of the 26th Conference of the Parties held in Glasgow, Scotland, in November 2021, seventy-four countries had adopted a national net-zero target as policy, although well over half of them had yet to enshrine it in law.[4] Even the United States belatedly got in on the act. President Joe Biden announced a goal of using only renewables for electricity by 2035 and achieving carbon neutrality by 2050.[5] Many US states, including California, New York, and Washington, have set mid-century net-zero carbon targets in law.[6] Of course, setting a goal is not the same as meeting it, and emissions have to start falling well before 2050.

Why are so many countries promising to get rid of carbon? Could clean energy sources alone – wind, sun, water, biomass, geothermal, and nuclear – do what coal, oil, and natural gas have been doing for well over a century? Is the goal of decarbonizing the global energy system this century at all realistic? And why are so many countries wanting to achieve that goal?

The *why* part of this commitment is straightforward: The planet is warming due to the accumulation of carbon dioxide and other greenhouse gases in the atmosphere. There is now a global consensus that the world is on its way to suffering harsh and unmanageable effects from this warming, including perhaps runaway climate change that would prove to be more disruptive and harmful than anything that has occurred in human history. At the same time, the range of clean energy technology options has expanded, with wind and solar prices falling and their capacities growing. The focus of this movement is on energy. Much also has to happen, of course, in agriculture, forestry, land use, and elsewhere – but global energy accounts for most emissions. If a transition to clean (zero or low-carbon) energy is not achieved, the battle to stabilize the climate will be lost.

If the *why* of this transition is clear, the questions of *whether* it will occur and *how* are less so. If the goal is to eliminate carbon emissions sometime in mid-century, or get them so low that remaining emissions can be offset by technological or natural means of carbon removal (hence the carbon neutrality), then this is a tall order, given that fossil fuels have made up over 80% of global energy for decades. Although countries are making progress on cleaning up their electricity sectors, nearly all of transport and much of industry still rely on fossil fuels. The outlines of a transition have taken shape, as we will see in this book, but the specifics are open to debate.

The Clean Energy Transition

The global transition to clean energy must occur because the energy system accounts for three-fourths of the greenhouse gas emissions that cause climate change. It also is the leading cause of health-damaging air pollution and many forms of ecological degradation. The transition will occur because it now is underway, and the forces driving it are compelling. The question is whether it will occur fast enough and in ways that meet social and economic goals.

It is not currently happening fast enough. Over four-fifths of the global energy system relies on fossil fuels like coal, oil, and natural gas. This has changed only marginally in decades, despite the threat of climate change. Moving to clean energy will require greater use of renewable sources – solar, onshore and offshore wind, geothermal, beneficial biomass, tidal, and wave – as well as major improvements in the efficiency and use of distributed sources like community or residential solar. It also means applying the fruits of modern technology to every aspect of the energy system: production and generation, distribution networks, storage, mobility, transport, industry processes, and electricity grids. All this will involve changing product designs, service delivery, business models, land-use practices, economic policy, and consumer behavior.

Because the transition is underway, there is room for optimism. Because there is so far to go, there is just as much room for pessimism. Indeed, energy offers a good news/bad news story. First, the good news:

- Between 2018 and 2050, capacities for generating electricity with solar photovoltaic (PV) sources could increase by a factor of twenty, while onshore wind may grow ten-fold.[7]
- The **energy intensity** (i.e. the energy needed to produce a dollar of economic output) of the global economy has improved by an average of 1.7% annually over the last two decades and is forecast to get even better over the next thirty years, with an average gain of 2.3% each year.[8]

- Solar PV and wind sources could be producing 62% of the world's electricity by 2050.[9]
- The costs of generating electricity from floating offshore wind platforms may fall by nearly 40% between 2019 and 2050.[10]
- The International Energy Agency (IEA) believes that "All the technologies needed to achieve the necessary deep cuts in global emissions by 2030 already exist."[11]

Then there is the bad news:

- By 2050, the world is likely to see population growth of 23% and average per capita income gains of 63%. Both types of increase have historically led to higher energy use.[12] Although good news for developing countries, this is bad news for the climate, health, and the environment.
- Despite the rapid growth of wind and solar, the International Renewable Energy Agency reports that "Energy-related CO2 emissions have risen by 1% per year over the last decade."[13] This was pre-pandemic, through 2019, but those growth rates will return.
- Despite the recent progress in renewables, the world is not on track to meet the goals set out in the 2015 Paris Agreement (discussed in Chapter 3).[14]
- To decarbonize global energy by 2050, annual investment should average *at least* $3.2 trillion a year to 2050; the recent average (2014 to 2018) was $1.8 trillion.[15]
- Vehicles have become more efficient, but gains "have largely been offset by trends toward larger vehicles."[16]

The Latest But Different Energy Transition

Energy is essential to the world as we know it. Prosperity only became possible almost two centuries ago because people learned to harness forces other than human or animal muscle for work.[17] Until the early 1800s and the dawn of the industrial revolution, humans relied on wood as their primary energy source, supplemented by wind and water. Early technologies like the steam engine, combined with the availability of coal as a fuel source, changed all that. Great Britain became the first industrial nation because it had coal, technical ingenuity, and an economy suited to applying energy to industry. The transition from wood to coal constituted an early energy transition. Since then, there have been transitions to oil, natural gas, nuclear, and now modern renewables like wind and solar. The long-term trends making up major European energy transitions are given in Figure 1.1.

For 200 years, energy transitions resulted from considerations of efficiency, cost, convenience, and need. Coal is a dense source of energy and is available in much of the world. Indeed, economic growth through the nineteenth century correlates nicely with coal's availability. Oil is energy-dense and portable, making it a hands-on choice for mobility – virtually all global transport is powered by oil. An early by-product of oil production, natural gas, became a heating and cooking fuel once pipeline technology enabled long-distance transmission. More recently, natural gas has edged out coal for generating electricity in many countries. Historically, countries used fossil fuels as a platform for achieving economic growth and more prosperous lifestyles. This must change.

The history of energy is one of transitions. Well into the 1800s, fuel for doing work beyond what human and animal muscle could

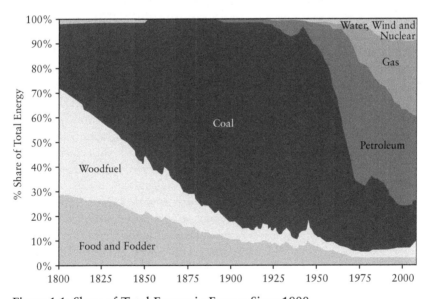

Figure 1.1: Share of Total Energy in Europe Since 1800
Source: Roger Fouquet, "Historical Energy Transitions, Speed, Prices, and System Transformation," *Energy Research and Social Science* 22 (2016).

do came from wood. When the revolutionaries and mutinous troops stormed the Bastille to start the French Revolution in 1789, coal was a novelty and oil a distant prospect. Thomas Jefferson likely wrote the Declaration of Independence using a lamp powered with whale oil. Coal largely replaced wood in the early to mid-1800s as the industrial revolution unfolded, first in England, then in Europe and North America. With their high energy density and portability, oil and other liquid fuels enabled major transport changes in the 1900s.

Coal, oil, and natural gas dominated the twentieth-century energy landscape. Major events in the history of that century are linked to energy production and use. Stokers were almost certainly busy filling the *Titanic*'s twenty-nine triple-furnace boilers with coal on the night of April 15, 1912, when it collided with an iceberg in the North Atlantic Ocean. What is seen as the start of mass production – the manufacture of Henry Ford's Model T from 1908 to 1927 – cemented oil's status as the fuel of choice for automobiles. Nuclear technology grew out of work on the atomic bomb during World War II, and smart electrical grid technologies are grounded in semiconductor design and production.

Roger Fouquet has written that energy transitions "depend on a series of actors and forces creating a new path."[18] They result from interactions between technology, perceived needs, human ingenuity, and events. The dominant energy industries in any era tend to lock-in and resist change, but new ones emerge to challenge the incumbent fuels and technologies. This is a slow and uncertain process. Vaclav Smil, a foremost expert on energy transitions, describes them as "prolonged affairs that take decades to accomplish." They "encompass the time that elapses between an introduction of a new primary energy source ... and its rise to claiming a substantial share (20 percent to 30 percent) of the market."[19] Even when governments push a transition, it does not happen overnight.

The world has now embarked on another energy transition. Unlike those of the past, this one is driven not just by economics, convenience, or new technologies. It is distinctive in being driven by worries over the social and environmental effects of energy production and use. Public opinion in most of the world is coalescing on the role fossil fuels play in climate change. But these climate worries are reinforced by other longstanding concerns. Fossil fuels are bad for human health. Burning them releases harmful pollutants like nitrogen oxides, sulfur dioxide, mercury, and fine particles that cause illness and premature death. Fossil fuels damage the environment by polluting water and land with acids, producing waste that ends up in rivers and lakes, and causing oil spills, among other harms.

Political and social concerns also shape the current transition. For critics, energy concentrates economic and political power in socially harmful, inequitable ways.[20] Until recently, oil companies were among the most prominent global corporations. The fossil fuel industry exercises an outsized influence in countries with fossil fuel resources, like the United States and Australia. Fossil fuels sustain authoritarian regimes in such places as Saudi Arabia, Iran, and Russia. At a deeper level, there are concerns about the centralization of political and economic power in relation to energy systems. The growth of electricity grids in the last century and the reliance on technologically complex, capital-intensive coal and nuclear power plants are part of what is driving this concern.

This chapter begins our look at the global energy system. As already suggested, there are grounds for both optimism and pessimism. From a technology and economic perspective, we can be hopeful. Many clean energy technologies already exist; others are in sight and can be realized with the right research, investment, incentives, and policy designs. The hard part is agreeing on the need for change, coordinating all the choices and actions that will be necessary, treating affected groups fairly, and taking the long view. As Albert Einstein said at a Princeton conference in 1946: "Politics is more difficult than physics."

The next section gives an overview of the global energy system. Obviously, this is a big topic, but the limited goal here is to provide a foundation for discussing decarbonization. Following that we take a look at concepts that are critical to understanding decarbonization and the forms it may take. Too often, clean energy debates reflect wishful thinking on one side and doom and gloom on the other. Clarifying the key concepts involved will provide a basis for defining a realistic path to a decarbonized global energy system. The last part of the chapter previews the book.

The Global Energy System

The concept of a system refers to a collection of interrelated parts that form a functioning whole in terms of both structure and process.

Systems exhibit a number of features. They are self-organizing to varying degrees in balancing external pressures with internal structure and processes to maintain equilibrium. They are resilient to the extent they maintain equilibrium despite these external pressures. Well-functioning systems adapt to pressures from the environment. Systems are or should be dynamic, regularly adapting to change as they strive to maintain their core structure and processes.

Thinking of global energy as a system helps to make sense of it. This system is complex and global. It has many parts: oil producers in Saudi Arabia; refineries in Texas; solar module factories in China; wind turbine firms in Denmark; public policymakers in California or Germany; farmers charging phones in Kenya; villagers cooking in India – the list goes on. The energy system is embedded in economic and social systems. It adapts to changes in demand and supply, government policy, new technologies, consumption patterns, the availability of fuels, and much more.

The energy system is made up of combined natural and social forces. Most energy comes from nature in one way or another: from fossil fuels, sun, wind, biomass, waves, and tides. The system adapts to pressures that are unpredictable, some coming as "shocks." Indeed, we speak of the 1970s oil embargo, when Saudi Arabia cut off oil exports to protest US support for Israel, as oil *shocks*. As with any shock, different parts of the energy system responded in contrasting ways: France turned to nuclear; Denmark began its transition to wind; Sweden later adopted a stringent carbon tax.[21] The United States responded similarly at first, but later reverted back to the fossil-fuel-driven status quo.

As the history of energy transitions shows, the global energy system has adapted over time. Government policy has influenced past energy transitions, providing subsidies, building infrastructure, promoting technologies, structuring investment incentives, regulating sources, and taxing energy. One theme of this book is that this transition requires a more active and intentional government role.

Energy basics

Learning about any policy field involves learning a language. This is true for energy. Some terms are so central to the clean energy transition that they are worth presenting and defining here. For example, primary energy comes from natural sources, from either *stocks* (coal, natural gas) or *flows* (wind or solar). Primary energy needs to be converted before it can be used. Crude oil is refined into gas or diesel; coal or wind are used in power plants to generate electricity. Two key terms are "total primary energy supply" and "total energy consumption." The latter is always smaller because of the losses incurred in converting and distributing primary energy. The heart of the challenge is matching supply with demand. The path to a transition lies in developing cleaner supplies while controlling or reducing consumption.

Electricity is "energy harnessed from the configuration or movement of electrons."[22] When we use electricity, we are using energy held temporarily by electrons. It converts primary energy from coal, natural gas, water, or wind into a usable form. It is not primary energy; it is a *carrier* of energy. Electricity makes up some 20% of global energy consumption. A goal of the transition is to electrify energy uses that depend on fossil fuels in transport, industry, and buildings. **Electrification** enables the use of clean technologies like wind and solar. Hydrogen is another energy carrier; as with electricity, primary energy is needed to produce it. Yet hydrogen does things electricity cannot (see Chapter 7).

Energy may be used more or less efficiently. Given likely economic and population growth, another central goal of the transition is to use less energy to produce a unit of economic output. The less the *energy intensity* of an economy, the more the return on each unit of energy used. The correlate of intensity is *productivity*: lower energy intensity equals higher productivity. Economic growth nearly always leads to more energy use but, at the same time, to declining intensity. In the US, intensity has fallen about 2% annually since 1974. In the same period, consumption grew by about a third. It would have grown far more without the fall in intensity. The same pattern occurs elsewhere.

One way to look at energy is through the United Nation's Sustainable Development Goals (SDGs). The relevant goal is SDG7, on "affordable and clean energy." Its three targets are to ensure "universal access to affordable, reliable, and modern energy services … increase substantially the share of renewable energy in the global energy mix … [and] double the rate of improvement in energy efficiency."[23] Electricity access is vital for overcoming poverty. In this respect, the goal is to increase access especially to electricity. This is improving. The number of people lacking access fell from 1.2

billion in 2010 to 789 million in 2018, with notable gains in India, Bangladesh, and Kenya. Ideally, people gain access to clean energy that is healthier, more climate-friendly, and accessible when they lack connections to bulk grids.

Global energy system highlights

Despite signs that an energy transition is underway, the last half century has seen little progress in getting the carbon out. In 1973, according to the IEA, 87% of the total global energy supply came from fossil fuels: coal, oil, and natural gas. In 2019, the comparable share was 81%. Fossil fuels had fallen only 6% in five decades.[24] Electricity generation is better. This is where the energy transition starts. In 1973, again based on IEA data, fossil fuels accounted for just over 76% of electricity generation; by 2019, this was down to 64%.[25] This is progress but it is hardly a roaring start to an energy transition. Tables 1.1 and 1.2 list the sources of total energy and just of electricity in 1973 and 2019.

The amount of energy delivered as electricity more than quadrupled between 1973 and 2019, from about 6,000 terawatt-hours (trillions of watts/TWh) to nearly 27,000. This trend will continue. Indeed, it must do so if wind and solar electricity is being used to decarbonize transport and industry.

Essential for understanding the energy system overall, and avenues for decarbonizing in particular, are regional changes in supply. In 1973, members of the Organisation for Economic Cooperation and Development (the OECD, a group of thirty-eight developed countries with relatively high living standards) accounted for 62% of the global

Table 1.1: Global Share of Total Energy Supply, 1973 and 2019 (%)

Source	1973	2019
Coal	24.7%	26.8%
Oil	46.2	30.9
Natural Gas	16.1	23.2
Nuclear	0.9	5.0
Hydro	1.8	2.5
Biofuels/Waste	10.2	9.4
Other	0.1	2.2

Source: IEA, *Key World Energy Statistics*, 2021, p. 6.

Table 1.2: Sources of Global Electricity Generation, 1973 and 2019 (%)

Source	1973	2019
Coal	38.9%	36.7%
Natural Gas	12.1	23.6
Oil	24.8	2.8
Nuclear	3.3	10.4
Hydro	20.9	15.7
Non-Hydro Renewables & Waste	0.6	10.8
Total (Terawatt-Hours/TWh)	6131 TWh	26,936 TWh

Source: IEA, *Key World Energy Statistics, 2021*, p. 30.

energy supply; China made up a mere 7% and the rest of non-OECD Asia just over 5%. In 2019, China's share had passed 23%, and the rest of non-OECD Asia was approaching 14%. Meanwhile, Africa, a next potential economic growth area, increased from a 3.3 to 5.9% share. The OECD share had fallen to 37% by 2019.[26] Energy supplies in OECD countries will be stable in the coming decades, but the expected rapid growth in Asia and eventually much of Africa will challenge clean energy ambitions. Fast-growing economies have big energy appetites.

A look at the dirtiest energy source – coal – gives a sense of the economics and politics of the energy transition. In 1973, OECD countries accounted for well over one-half (56%) of global coal production. China and the rest of non-OECD Asia were under 18%. In 2019, China made up half (49.7%) of coal production and the rest of non-OECD Asia another 19.6%. OECD countries had fallen to one-fifth. Among the other countries with large coal production shares are India (10% of world production), Indonesia (8%), the US (8%), and Australia (6%).[27] The largest coal-exporting countries – including Indonesia, Australia, Russia, South Africa, and the US – have economic reasons for resisting clean energy policies and investments. The same holds for oil producers and exporters. The US (17%), Russia (12%), Saudi Arabia (12%), and Canada (6%) produce nearly half the world's oil. The largest exporters are Saudi Arabia, Russia, Iraq, Canada, and Nigeria. Among the major importers are China, the US, India, South Korea, and Japan.[28]

As Table 1.1 shows, coal rose slightly in its share of the total energy supply, and natural gas even more. Oil declined in supply share, largely because it is used less to produce electricity. Hydro grew; biofuels fell slightly. The "other" category reflects wind and solar growth. Still, even with their rapid growth rates, modern renewables are a small fraction of energy. Meanwhile, energy supplies over the 1973–2019 period more than doubled. The growth in clean energy was overwhelmed by higher energy use as a result of more manufacturing and more prosperous lifestyles.

The US Energy Information Administration (EIA), part of the Department of Energy, makes an annual assessment of energy trends and prospects for the next several decades. Especially useful is its *Reference case*, which incorporates likely trends, planned infrastructure change, incremental technology gains, and projected costs into 2050. It does not account for unanticipated events, like changes in international agreements, disruptive events, technology breakthroughs, or unannounced law and policy changes. For the period 2020–50, it projected a most likely global economic growth rate of 2.8%. The rate will vary by stage of development. In the OECD countries, the likely growth rate is 1.7% (a range of 1.2–2.2%), while in non-OECD countries it is 3.6% (a range of 2.6–4.6%).[29]

Even with gains in energy intensity, this growth will burden the climate and threaten public health. The EIA projects that global energy use will grow by nearly half by 2050, "driven by non-OECD growth and population."[30] Indeed, GDP is projected to nearly double in OECD and almost quadruple in non-OECD countries. Population is expected to remain level in the OECD but to increase outside of it. In the energy mix, the EIA expects coal and nuclear to remain flat, oil and natural gas to grow, and renewables to nearly double to become 30% of the total, driven by wind and solar.[31]

Per capita energy consumption varies greatly across countries. This mostly reflects the size of an economy and its wealth, but not entirely. Generally, richer countries like the US or Japan use far more energy per capita than poor ones in Asia or Africa. Yet countries at similar wealth levels also vary. Among high-income countries, the US in 2019 used 288 gigajoules per capita, Canada 380, and Australia 254. Germany's per capita use was 157 and the United Kingdom's 116. China was 99 per capita, Vietnam 43, India 25, and Bangladesh 11.[32] As economies grow, they use more energy, but at varying rates. This has to change if the world is going to decarbonize.

The good news is that as energy use increases, energy intensity improves. Lower intensity means that an economy or sector (chemicals, auto, iron and steel) generates more economic value per unit of energy use. Economies become more energy efficient as they develop and technologies improve. In 1990, the world economy needed the equivalent of 2.11 kWh hours of electricity to produce a US dollar of output; by 2015 this had fallen to 1.43 kWh. Energy intensity improves even with advanced growth. High-income economies like the US and EU stood at 1.82 per kWh in 1990 but had fallen to 1.28 by 2015. Low-income countries improved at the same rate (30%), although starting higher, and middle-income countries at a higher rate (see Table 1.3).[33] Still, to decarbonize, *absolute energy use and not just intensity must fall.* Given the expected economic growth rates, global energy intensity has to improve at roughly double the rate of recent years to come close to decarbonizing.

Another way to define the landscape, especially given a decarbonization goal, is by source of emissions. According to the WRI's *ClimateWatch*, global greenhouse gas emissions in 2018 (including land use changes and forestry) were just under 49 gigatons (billion tons) of carbon dioxide equivalent (CO_2e), some 50% above the 1990 figure – not a good trend.[34] Energy accounted for three-fourths (76%) of this. Among countries, the top ten accounted for over 60% of CO_2e emissions (for the purposes of this book the EU is counted as one "country"). China was 24% of the total; the US was 12%; India and the EU each were some 7%; and Russia was 4%.[35] If these countries are not part of the global decarbonization process it will not succeed. If we exclude the effects of changes in land use and forestry (LUCF), global emissions amount to just under 47 gigatons, of which a similar percentage (77%) are energy-related. Excluding LUCF, global emissions per capita in 2018 were 6.26 tons;

Table 1.3: Energy Intensity of Economies, by Income Group, 1990 to 2015 (kWh/$)

Income Group	1990 Energy Intensity	2015 Energy Intensity	% Improvement
Low Income	3.35	2.46	30%
Middle Income	2.56	1.54	40%
High Income	1.82	1.28	30%
World	2.11	1.43	32%

Source: *Our World in Data*, "Energy Intensity of Economies, 1990 to 2015" (World Bank data).

global emissions of all greenhouse gases per million dollars of GDP amounted to about 550 tons.

Countries vary considerably in emissions per capita and per unit of economic activity. Table 1.4 compares selected countries at various stages of economic growth. For each, it lists the total 2018 emissions, the energy-related percentage, and those per capita and per million dollars of GDP. Countries vary in the proportion of emissions that are energy-related. The percentage of energy-related emissions is lower for countries earlier in their growth process, such as India (72%), Nigeria (63%), and Kenya (35%). As they grow, the share of energy-related emissions will increase, especially if they rely on fossil fuels. As noted in the table, these emissions numbers exclude the effects of land use changes and forestry (LUCF).

Countries also vary in emissions per capita, from a high on this list of over 18 tons in the US, to a mid-range for Japan and the EU, to lows for India and Nigeria. With expected economic growth in coming decades, the last column in Table 1.4 is notable. Although the mature economies (US, Japan, the EU) have high per capita emissions, they are low per unit of GDP. Their economies are less carbon-intensive than in early-growth countries. Still, with economic

Table 1.4: Emission Profiles of Selected Countries, 2018

Country	Greenhouse Gas Emissions (CO2e gigatons)	% Related to Energy	Per Capita Emissions (tons)	Tons / Million $ GDP
China	12.36	83%	8.87	889.2
United States	6.24	84%	18.84	292.7
India	3.37	72%	2.50	1243.9
EU	3.57	81%	7.98	223.4
Russia	2.54	90%	17.60	1523.4
Indonesia	.970	62%	3.62	930.3
Japan	1.19	92%	9.38	239.5
United Kingdom	0.452	81%	6.80	127.5
Nigeria	0.311	63%	1.59	784.1
Kenya	0.079	35%	1.53	898
Sweden	0.046	78%	4.56	83.5

Note: Emissions totals exclude land use change and forestry (LUCF) to present a country's direct impact on global emissions.
Source: Based on data from WRI, *Climate Watch*, https://www.climatewatchdata.org./ghg-emissions?source=CAIT.

growth, emissions will grow, and that is a problem. Sweden is impressive, with low emissions per capita and per unit of GDP. Sweden shows that affluence is not incompatible with mitigation; it is no surprise that Sweden ranked at the top of the 2020 *Climate Change Performance Index.*

Another way to look at countries is in terms of their cumulative emissions over time.[36] Between 1751 and 2017, the world emitted over 1.5 trillion tons of CO2. The US accounted for 25% of that, followed by China at nearly 13%. Together, the EU-28 are responsible for 22% of historical emissions. The South American and African continents account for 3% each. India – the expected rapid-growth economy in the coming decades – is likewise so far responsible for a mere 3% of cumulative emissions.

The coming decades will have to be different from the last five. True, decarbonization did not even emerge on global agendas until the 1992 UN Framework Convention on Climate Change. Even since 1990, however, fossil fuel shares have fallen only slightly. By late in the century, fossil fuels will have to have fallen *at least* to 20% and *ideally* to under 10% of total energy, with any residual emissions offset by enhanced natural or technological carbon removal. Otherwise, there is little chance of reaching the holy grail of carbon neutrality.

The big energy picture

As Jonathan Harris and Brian Roach put it in their text on natural resource economics: "The period of intensive fossil fuel use that began with coal in the eighteenth century was a one-time, unrepeatable bonanza – the rapid exploitation of a limited stock of high-quality resources, with increasingly negative effects on planetary ecosystems."[37] The planet will endure, but the ability of humans to live a comfortable, safe, and meaningful life is under threat. To summarize the issues raised so far:

• The global energy system accounts for some three-fourths of climate-harmful emissions and must be a priority in the fight against climate change.[38]
• The trajectory of rising emissions has slowed but not changed (excepting the 2020 pandemic year, which bought the world some time but did not alter the trajectory).

- Poor countries face different issues. They need to expand electricity access to enable growth and improve the quality of life, ideally with renewables.
- Per capita consumption and emissions vary among countries. But energy use and associated emissions rise with economic growth, despite better energy intensity.
- Early industrializers (US, Europe, Japan) accounted for most emissions until now, but emerging, rapid-growth countries (China, India, Indonesia, Vietnam, Nigeria) will add the largest share of energy-related emissions for the rest of this century.
- Energy intensity gets better as economies mature; this must translate into absolute cuts in energy use and cleaner sources, leading to far lower **carbon intensity**.
- Energy's future is in electricity and eventually green hydrogen; electricity's future is in wind and solar with support from complementary technologies like nuclear or carbon capture and storage.
- Transport, buildings, and heavy industry are the most difficult sectors in terms of energy use. Nothing is easy in the path to clean energy, but these are the toughest nuts to crack.

One reason the world is having a hard time getting off fossil fuels is *the tremendous appetite for energy*. Fossil fuels made prosperity possible. Since 1960, the global economy has grown, with exceptions, at a rate of about 2–6% a year.[39] The global population in 2021 was 7.5 billion, and will exceed 9 billion by the 2050s. Politically and economically, reliable energy is not optional. A second reason is **carbon lock-in** – the extensive investment of technology, physical capital, and social familiarity in current energy systems, along with the political power of the fossil fuel industries. Third is *the complexity of the transition*. For example, the rapidly falling costs of solar photovoltaic systems are but one piece in a complicated puzzle. Grids have to be designed and managed to integrate variable renewables; transport has to be electrified; surplus wind and solar energy has to be stored. Energy systems are complicated, their parts interdependent.

The clean energy transition is often oversimplified. We hear calls for 100% renewables or for shutting down coal and nuclear plants without an appreciation of just what that involves. A transition built on false premises or unrealistic expectations is one that almost certainly will fail.

Making Sense of Energy: Key Concepts

The energy landscape offers several concepts for defining choices, comparing and evaluating alternatives, and proposing solutions. They are a starting point for making sense of the paths to clean energy.

Energy systems involve lots of physical infrastructure, depend heavily on accumulated past choices, and are embedded in economic and social systems. In describing energy systems as complex, we mean not just that they are technically intricate (try explaining light-water nuclear reactors or chemically based energy storage) but also that they involve an interdependence among their various parts. Wind and sun help only if their power can be converted into useful energy and distributed to users. To get the full benefit of electric vehicles (EVs), they need to be charged by electricity generated from renewables. Energy systems are intricate spiderwebs; touching one part affects many others.

The reliance on physical infrastructure is obvious to anyone who has seen electricity substations or gas stations on busy roads, or the huge cooling towers at coal-fired and nuclear power plants. These physical structures represent huge financial investments. The effects of past choices, what social scientists call *path dependence*, are apparent. Decisions on nuclear technology made in the 1950s shaped the design of the more than ninety reactors operating today in the US.[40] The increase in average global temperatures that is occurring was influenced by carbon dioxide released since the start of the industrial revolution and from Henry Ford's Model T. Greenhouse gases emitted and carbon sinks lost today will have effects for generations. We need to look forward as well as backward.

As for energy systems being embedded in social and economic systems, think of how a world without modern energy would look. It would be poorer, to be sure. High-income countries use lots of energy.[41] A result mainly of fossil fuels, emissions move almost in lock-step with GDP. Studies have found that "energy indices are highly correlated with a higher standard of living,"[42] and that "indices of human social well-being are well correlated with indices of energy availability."[43] The link between economic growth and CO2 emissions can be broken – a break known as *decoupling* – but growth as it has occurred to this point has almost always involved higher emissions due to the use of more fossil fuels.

Carbon lock-in

Our look at the key terms begins with *carbon lock-in*, which refers to the dependence on carbon-based sources resulting from decades of accumulated choices. It "arises from systematic interactions among technologies and institutions."[44] Technologies emerge for productively using a source, say coal or oil. Institutions are created to support and expand energy technologies and their associated infrastructures. Governments adopt policies to facilitate the expansion of fuel sources and technologies, often through subsidies.

Two examples of carbon lock-in are the array of infrastructures and institutions created for generating and distributing electricity and for supporting mobility with internal combustion engines. Electricity grids in developed countries are elaborate expressions of human ingenuity and design. They apply engineering and management prowess to deliver electricity to billions of people – on demand and mostly reliably. In a history of the US grid, the anthropologist Gretchen Bakke describes it as "a complex and expansive electrical delivery system that we care little for and think even less about."[45] Through choices made early in the last century, the bulk grid evolved into a centralized service in which large utilities linked power generators to users via elaborate distribution systems. The grid historically has relied on electricity generated with coal, and later nuclear and natural gas. Although changing, the grid still is complex, carbon-based, and supported by a formidable physical and institutional infrastructure.

The technical and institutional complex built up around the internal combustion engine (ICE) is equally impressive. Oil is to mobility what coal was to electricity, although more so. Based on convenience and reliability, the ICE powered by oil won out as the transport technology of choice at the start of the twentieth century. A vast and complicated production, supply, processing, and distribution chain grew up around the ICE, one that is truly global – from drilling and refining, to continental and global scales of transport, to mass distribution at fuel stations. The ICE shaped modern society, determining land use, housing, and lifestyles as well as transport. Public policy in the US especially supported the ICE complex, with average federal subsidies of nearly $5 billion (2010) annually from 1918–2009.[46] The US interstate highway system alone cost some $500 billion in recent (2016) dollars.[47] It is no wonder that carbon is locked in.

Carbon lock-in represents a great deal of physical and institutional infrastructure, investment, sunk costs, and socio-economic relationships. It also represents two other characteristics that shape the clean energy transition: distributions of political power and consumer familiarity. Carbon lock-in is as much as anything a statement about who has political power in society.[48] In general, the larger and more influential the fossil fuel interests in a country are, the tougher the clean energy transition will be.[49] The politics of carbon lock-in are a formidable barrier to change. Another barrier is what Gregory Unruh calls *adaptive expectations*, referring to consumer confidence in the quality, performance, and permanence of new technologies.[50] Put simply, we want to be comfortable with technologies and confident about their convenience and reliability.

Capacity factors

This term applies to technologies for generating electricity. A pillar of decarbonization is that we must electrify transport and industrial activities that currently rely directly on fossil fuels. The strengths and weaknesses of various methods for generating electricity thus are critical issues.

A **capacity factor** is the ratio of actual to potential electricity available from a technology. An advantage of *dispatchable* sources like coal, nuclear, and natural gas is their high capacity factor, which typically falls within a range of 80–90%. Generating plants shut down for periods for maintenance or due to malfunction, but the fuel is on hand and generates electricity steadily and reliably. Coal and gas are dispatchable; wind and sun are not. Among renewables, geothermal does well, with capacity factors comparable to dispatchable fuels. Hydro is next with capacity factors ranging from 30–60%, depending on size and location. Wind and solar, sources that will power the clean energy transition, have lower capacity factors. Onshore wind is 30–55%; offshore may be higher in good locations. Utility-scale solar PV falls into the 30–40% range; residential PV is lower (around 20%); concentrated solar with storage does better. These figures will improve, but they are a reality and a constraint.[51]

The difference between *dispatchable* and *renewable* fuels is that the latter are variable. The sun is an ideal energy source, but it is not available for much of any 24-hour period and varies by season and location. Wind is more reliable, hence its higher capacity factor, but it is also variable. Lower capacity factors present two challenges. One is cost: less electricity from **variable sources** translates into higher costs per unit of electricity and lower investor returns. Another is managing a reliable grid with renewables, or the challenge of **grid integration**. With their lower capacity factors, wind and solar are less available and dependable than fossil fuels or nuclear, which offer consistent *baseload* power. Indeed, grid integration is a major challenge, as will be discussed in chapters 5 and 6.

Energy returned on energy invested (EROEI)

Another concept is the energy gained from a technology or fuel compared with the energy needed to make it usable, or the *energy returned on energy invested*.[52] Many experts question the value of sources with a ratio of less than three to one, meaning that each unit of energy invested yields fewer than three units of useful energy. An historical advantage for fossil fuels is high EROEI, although that is falling as they become harder to exploit. A 2014 survey of research on EROEI found hydropower to be highest, at a ratio of 84:1. Given that the best hydro sources have already been developed, that will fall. For oil the ratio was 20:1, down from 30:1 in the 1990s. It is lower for sources like hard-to-develop Canadian tar sands, at 4:1. As exploration and drilling become more difficult, "the EROI of oil and gas will continue to decline over the coming decades."[53]

Wind and solar PV do not do as well. Wind *may* offer a reasonably good ratio of roughly 20:1, but PV is lower. Because they are variable, there is need for backup storage, and that reduces the ratios even further. This does not mean that wind and solar are not mainstays of a clean energy system; they are. But their lower EROEIs do have practical cost implications and other consequences.

Levelized cost of energy (LCOE)

Energy is a commodity. What counts in markets is its usefulness in final, delivered form, not where or how it is produced. Energy is pervasive in modern life: we depend on having reliable flows of energy to do work, provide light, deliver heating, power appliances, preserve food, and move people and goods. Its price affects nearly all aspects of modern life. Until recently, wind and solar costs were high, but with rapidly falling costs their weakness has turned into a strength.

The **levelized cost of electricity** is a tool for comparing energy technologies and sources. I focus on electricity given the need to electrify industry and transport with renewables. LCOE is the total life-cycle cost of delivering a kilowatt-hour (kWh) of electricity. It accounts for initial capital expenses and the costs of fuel, operation, maintenance, and decommissioning. Calculations usually assume a long (twenty to forty years) time frame. More on calculating LCOE is given in Box 1.1.

Two aspects of LCOE are critical for clean energy. First, the costs of wind and solar PV have fallen, especially the latter. By 2020, the LCOE for onshore wind and PV was competitive with coal and gas and improving. In energy terms, wind and solar are at *price parity* or better.[54] Offshore wind's LCOE was higher but will fall with experience, technology, and the anticipated scale-up.[55]

Why have costs fallen so fast? Chapter 5 examines wind and solar in detail, but the short answer is that the fall is due to technology innovation and **economies of scale**.[56] Wind and solar technologies are getting better. For wind, the gains have come from new turbine design and materials; for solar, from better conversion efficiencies and materials. Economies of scale describe a positive feedback loop: increasing volume lowers the costs of producing each additional unit. Indeed, Germany performed a global service by adopting generous feed-in tariffs (discussed in Chapter 7) that, in effect, subsidized an expansion of PV technologies and created a public good for the rest of the world.

The cost makeup for electricity sources varies quite a bit. For wind and solar, fuel costs are zero. Life-cycle costs flow mainly from the initial capital investment, which can make up 75% or more of the final electricity bill. The same holds for nuclear: high capital costs but minimal fuel costs. For coal and gas, fuel is a big part of the LCOE. Once wind or solar sites are running, the marginal costs of generating a unit of electricity are low, a benefit of "the low costs of turning wind turbines or harvesting sunlight."[57] This was an advantage during the Covid-19 pandemic.[58] Lower electricity demand, combined with low marginal prices, caused utilities to buy more wind and solar.

What tools like LCOE do not account for are the **social costs** of technologies and fuels. The climate, health, and ecological costs that result from the production and use of fossil fuels are not measured in market calculations. LCOE is a vital tool, but it does not give a full

Box 1.1: The Levelized Cost of Electricity (LCOE)

The LCOE is "an economic measure used to compare the lifetime costs of generating electricity across various generation technologies."[59] It includes the *capital costs* of building the plant, including financing costs; *operation and maintenance* (O&M) costs in the plant's useful life; and the *disposition costs* of decommissioning the plant at the end of its life. O&M costs are further broken down into *fixed* costs that do not vary with the amount of electricity produced (core staffing, insurance) and *variable* costs that do vary, such as costs of inputs (fuel costs).

Because these costs are incurred over a period of time, they are normalized to a net present value and converted to current dollars. An advantage of renewables like wind and solar is not having input (fuel) costs. They also tend to have lower disposition costs, especially compared to a nuclear plant at end of life. Fossil fuels involve input costs that vary over time. At the same time, renewable sources are not always available – they are intermittent or variable. This variability affects the capacity factor of a technology. Other things being equal, the higher the capacity factor, the lower the LCOE. Even with lower capacity factors, however, wind and solar technologies are increasingly competitive with firm, dispatchable sources like coal, natural gas, and nuclear. Chapter 5 gives more on LCOEs for renewables, most of which are declining.

picture of the non-market harm from fossil fuels, of the social costs, which are the topic of Chapter 2.

Five Themes for the Energy Transition

This book offers a point of view on the clean energy transition, aiming not to answer all questions about decarbonization but to highlight the options for and barriers to getting there. The book has five themes: clean energy is not optional; government must play a role; it is about technology but also much more; scale is everything; all zero-carbon options should be on the table.

Clean energy is not optional

That there should be a transition to clean energy in this century is not a matter of debate. Climate change is a major reason for this – a ticking time bomb that erupts not in one explosion but unfolds steadily with devastating certainty. We may not know *precisely* how much sea levels will rise, or how fast glaciers will melt, or just how many hurricanes or wildfires the world must endure, but all are happening and will grow worse so long as fossil fuels dominate global energy. But climate is only a starting point. Air pollution is the largest environmental health threat, and fossil fuels cause ecological harm. Clean energy is a path to a better world.

Government must play a role

This is the first significant energy transition in history driven primarily by social costs. Since markets do not account for these costs, government and public policy have to fill the gap. A purely market-driven transition will take too long even if it occurs as a result of market forces, such as steadily declining wind and solar costs or the eventual scarcity of fossil fuels. Relying on markets and scattered policy interventions without clear goals is a recipe for failure. Governments have to accelerate change with coherent policies, focused investment strategies, and consistent political commitment.

This is a social as well as technology transition

As will become clear in chapters 4–7, technology underpins this transition. Indeed, recent advances such as in solar PV and battery storage brighten the prospects for clean energy. Although there is much to be said for lifestyles that are less energy- and materials-intensive, the world depends on energy. Yet the social aspects of the transition are as important as the technology. For ethical and practical reasons, the transition should be equitable.

Scale is the operative word

If we follow the news, it seems there is always some new energy technology or solution that will save the day. Some believe the future lies with nuclear energy from fusion technology. Others place their stock in bioenergy with carbon capture and storage, which removes carbon from the atmosphere. Some visionaries foresee a world where platforms in space generate solar power at no cost to an energy-starved world. The floating space platform may be well off into the future, but technologies like bioenergy with carbon capture and storage (BECCS) exist now. The catch is that any technology for running a global energy system has to operate on a large scale. Scale does not imply universality; geothermal or tidal sources could meet special needs in limited situations. But it does mean that technology has to be available when and where it is needed at competitive prices.

All reasonable options should be on the table

There are few easy choices when it comes to meeting the world's energy needs. Some options are flawed and should be abandoned. Coal powered the industrial revolution but is the dirtiest, least healthy way to generate electricity. Oil is portable and energy-dense; it serves many purposes but is harmful in many ways. Even natural gas, the *cleaner* fossil fuel, affects the climate, ecosystems, and health, not only in carbon dioxide emissions but in methane emissions too. What of nuclear power, which is low-carbon but poses other risks? What of carbon capture and removal technologies? They enable some low-carbon fossil fuel uses and may remove carbon from the atmosphere. The approach in this book is to keep all feasible options on the decarbonization table and let the reader decide.

A Punctuated Equilibrium Framework for the Energy Transition

This is a public policy book. Its goal is to enable readers to understand the need for clean energy, to evaluate the forms it may take, and to assess the barriers that have to be overcome. In dealing with public policy issues, it often helps to have a framework for making sense of it all.

There are technological, economic, social, and political dimensions to the energy transition. One of my themes is that government and public policy must play a central role, which underscores the importance of politics. After all, politics is how a society's collective goals are determined and realized, and the politics of the energy transition will determine the form and pace in which it occurs.

In assessing the politics, I draw on a model known in policy studies as the *punctuated equilibrium* framework for explaining disruptive change. This was developed by Bryan Jones and Frank Baumgartner to explain why, after periods of stability and continuity, political relationships break down, new ones emerge, and policy change occurs.[60] This has been applied to many issues, including nuclear energy, smoking, and the environment, among others. It is a useful framework for thinking about the transition from a fossil-fuel-based global energy system, with its high social costs and concentrations of economic power, to a clean energy system founded on the pillars of efficiency, clean renewables, electrification, and complementary technologies like nuclear, carbon capture, and green hydrogen.

The punctuated equilibrium framework was in large part a reaction to the reliance in policy studies on *incrementalism*. The theme of incrementalism is that policy change occurs in small steps, builds on past decisions, involves negotiated agreement among actors, and leads to marginal rather than fundamental policy change.[61] This is a useful framework, but it tells us little about cases of disruptive policy change. In the 1970s, for example, countries like the United States, Great Britain, and Japan achieved substantial changes in pollution-control policies, to the extent that this period has been termed the *environmental decade*. Many forces converged to stimulate non-incremental change.

The punctuated equilibrium framework accepts that most policy change fits the incremental model and does not fundamentally disrupt the status quo. This is because specialized *political subsystems* hold power over sets of policy issues and are able to contain the scope of change. These subsystems benefit from the status quo; they thus seek to sustain autonomy, limit the scope of policy disputes, and protect a favorable (from their perspective) situation against disruptive change. At times, external forces challenge this control. When this occurs, the scope of relevant actors aiming to change policy expands, previously positive issue framings turn negative, and conflicts increase.

Clean energy policy reflects this pattern. The fossil-fuel-based energy system and the interests it represents, both within countries and globally, have long dominated energy policy. Until the 1970s oil embargos, public policy in nearly all countries focused on expanding energy supplies while managing such issues as safety and the monopoly power of utilities. Public policies were largely *distributive* – aimed at distributing benefits to existing industry actors – rather than *regulatory* – constraining behavior to achieve social goals. This began to change with the oil embargos. Still, what is now changing energy policy – making it more regulatory and less distributive – is the rise of climate change on the global agenda.[62]

Using a punctuated equilibrium framework, we can say that, in most countries and globally, energy policy was focused on maintaining supplies and controlling prices. Energy served to promote economic development. Fossil fuels and economic growth were closely linked. Energy policy was largely controlled by and benefitted fossil fuel interests, and later the nuclear industry. Government intervened when prices rose rapidly or shortages caused problems, as in the 1970s, but fossil fuel interests made up the political subsystems that could maintain the status quo and limit the scope of policy change. This is the political dimension of the phenomenon of carbon lock-in.

Worries about climate change disrupted the energy status quo. As the scope of political conflict expanded, energy policy was no longer the province of a narrow group of interests and institutions. The politics of climate change, reinforced by other social costs, are characterized by conflict, challenges to fossil fuel industry power, and pressure for a redesigning of the energy system. Energy politics have become more explicitly regulatory. Although still powerful, fossil fuel interests now have to contend with disruptive policies and pressures coming from the demand for clean energy.

The punctuated equilibrium framework is useful for thinking about the politics of the clean energy transition. Aside from the technological, economic, and social issues examined in this book, the politics of realizing a post-carbon world will be the critical factor in its success. I return to this framework in the final chapter by assessing the prospects for a decarbonized world. The reality is that an incremental approach to change will not do the job. The political challenge is to punctuate the status quo of the systemic carbon lock-in and realize a new equilibrium based on clean energy.

The Rest of the Book

The good news is that the world is engaged in a clean energy transition. The bad news is that it is not moving fast enough. The pace must be accelerated. This is the first energy transition driven by worries about social costs. These costs have not been ignored totally, to be sure. Much of modern environmental policy has focused on managing the social costs of fossil fuels, mostly health-damaging pollution, through clean-air laws and regulation. Until recently, the goal was not to transform energy systems but to limit the damage using technologies like catalytic converters in cars or devices for removing sulfur oxides from coal emissions. For carbon dioxide and other climate emissions, technology controls offer limited options. Hence, our focus has turned to an energy transition.

The next chapter examines the clean energy imperative – the need to move from a primarily brown to a primarily green energy system. Although climate change is driving this, the need to avoid the health and ecological harms of fossil fuels strengthens the case. But beyond the climate, health, and ecological damages of the existing energy system there are also opportunities, in terms of durable jobs, economic prosperity, social equity, a better quality of life, and even expectations of better governance.

What form will the clean energy transition take? There are options. One scenario envisages an energy system built entirely on renewables. Here, fossil fuels are gone, and the world runs on wind, sun, and water. In such a totally decarbonized world, technologies for capturing, sorting, and reusing carbon from residual uses of fossil fuels are seen as costly and unreliable, and nuclear power as too risky. Other scenarios assume that fossil fuels will still play a role. The goal here is to decarbonize, with renewables doing the heavy lifting, but to incorporate options like **advanced nuclear** and carbon capture and removal. Chapter 3 examines the big-picture scenarios for decarbonizing energy by mid-century.

Chapters 4 through 7 break these scenarios down. Chapter 4 considers the role of energy efficiency, or getting what we need from energy while using less of it. All scenarios rely heavily on improved efficiency. We cannot use more energy in order to decarbonize. This is a classic no-brainer: *saving energy also saves money*. The world has not taken full advantage of this not entirely free but relatively low-cost lunch.

Chapter 5 turns to the workhorses of the transition: wind and solar. Other renewable fuels and technologies also enter the picture – hydropower, geothermal, biomass, and ocean energy. Hydro and geothermal play a role but are unlikely to expand greatly beyond their current levels. Biomass encompasses a range of sources, many having worrisome negatives. Wind and solar made up a fraction of the total energy supply in 2021. This has to increase dramatically if global decarbonization goals are to be met. Although wind and solar are expanding rapidly, the pace of that growth has to accelerate.

Chapter 6 discusses the third element in decarbonizing: electrify everything, or nearly everything. Of all the parts of the energy system, electricity generation offers the best prospect for renewables. The International Renewable Energy Agency reported that renewables' share in electricity stood at 26% in 2018. Transport and industry are less encouraging. The IRENA's transformation scenario (Chapter 3) envisions the share of energy use from electricity growing from one-fifth to one-half or more by 2050; the number of electric vehicles must grow to well over a billion.

Chapter 7 focuses on contested issues in clean energy debates: nuclear energy and carbon removal and storage technologies. Nuclear is contentious due to safety issues, capital costs, security, and waste concerns. Major events – Three Mile Island (1978), Chernobyl (1986), Fukushima (2011) – have reinforced doubts in many countries about nuclear, despite its low-carbon character. Similarly, despite an acceptance of carbon removal and storage in decarbonization planning, many experts doubt its economic feasibility and question its continued reliance on carbon-based fuels. Still, low- to zero-carbon technologies are in the picture, given the challenges of carbon lock-in, stranded assets, and economic growth. As a source of reliable baseload power, nuclear helps integrate wind and solar into grids. This chapter also examines *green hydrogen*. Hydrogen has valuable clean energy qualities if it is produced from renewable sources.

Government, as has been said, must play a central role in the transition. Chapter 8 presents the policy tools governments may use for a transition. Public policy must facilitate each part of the process: overcoming carbon lock-in; investing in research; accelerating renewable ramp-ups; delivering efficiency tools; creating investment incentives; promoting end-use electrification; building flexible, technology-**smart grids**; ensuring an equitable transition; and generally guiding the path to clean energy.

The final chapter focuses on the political and social dimensions of clean energy and the prospects for a decarbonized world. Technology is critical, but the transition is about more than technology. It is also a social transition that should be just and fair. Critics of the existing energy system argue that it promotes economic and political inequity. Without doubt, some groups bear the costs more than others, particularly in terms of air pollution and the effects of climate change. **Energy democracy** advocates go further, arguing that the current energy system reinforces existing concentrations of political power. In closing, we consider the prospects for the clean energy transition. Can it be accomplished at the needed scale and in time? And how might it look?

Final Points

Many definitions of sustainability exist. They share a concern for the future, a recognition of planetary limits and a commitment to respecting them, and a high regard for equity and human dignity. Clean energy underpins all of this. Most obvious is respect for planetary limits; any response to climate change has to include a shift away from fossil fuels. A healthy planet and global population depend on our ability to meet energy needs in clean, creative, and efficient ways.

This book was written during the 2020–21 coronavirus pandemic, which profoundly affected all aspects of modern life, including energy. In 2020, global fossil fuel emissions fell by some 7% relative to 2019, mostly from lower transport emissions.[63] Numbers for clean energy investment and jobs also fell, but wind and solar still grew. The decline in emissions gave the world breathing space, to be sure. Still, if the fundamentals do not change, this will have been only a respite. Long-term progress depends on new investment, policies, and cooperation. The pandemic's effect is not "its growth impact but its influence on national commitments to action."[64] Emissions began to grow again in 2021.[65]

The Russian invasion of Ukraine in late February 2022 demonstrates how events can upend plans for an energy or any other kind of global transition. This development in particular has implications for clean energy; Russia is a major natural gas and oil exporter, exceeded only by the US and Saudi Arabia. Europe depends heavily on Russia, the source of 40% of its natural gas and one-fourth of its oil.[66] Economic sanctions against Russia will include energy cuts, which removes a major income source for the Russian economy. In Europe, the Ukraine invasion is likely to speed up the clean energy transition, as countries seek to offset their dependence on Russian gas and oil. The effects in the US are difficult to predict and will depend on which governing coalition determines energy policies. It will either accelerate the transition to renewables or cause the US to double-down on domestic fossil fuels. The big question concerns China, India, and other Asian countries, where most future energy demand will occur. If Russia succeeds in expanding exports to Asia, this may have the effect of slowing down the clean energy transition in places where it will be most needed.

The Ukraine invasion constitutes one of the most significant challenges to global security since the Second World War. It also illustrates the uncertainty that affects major global transitions. Indeed, the fear is that climate change could "be relegated by world leaders to an afterthought."[67]

My theme is decarbonizing – getting carbon out of the global energy system. To be sure, the social costs of energy involve more than carbon dioxide. Other pollutants – methane, nitrous oxides, specialty gases – also cause climate change. Public policies for these are necessary, but they are not the main issue discussed here. If we succeed in building a clean energy system by mid-century, we will mitigate the leading causes of climate change and eliminate damaging sources of air pollution. The next chapter starts with the reasons for accelerating the transition to clean energy.

Guide to Further Reading

Richard Rhodes, *Energy: A Human History*, Simon & Schuster, 2018.

Walter A. Rosenbaum, *American Energy: The Politics of 21st Century Policy*, Sage/CQ, 2015.

Vaclav Smil, *Energy Transitions: History, Requirements, Prospects*, Praeger, 2010.

Gregory Unruh, "Understanding Carbon Lock-In," *Energy Policy* 20 (2002), 817–36.

2

Why Clean Energy Matters

Readers of nineteenth-century English novels have an idea of how bad London's air could be at the time. In *Our Mutual Friend*, his last novel, Charles Dickens wrote: "It was a foggy day in London, and the fog was deep and dark. Animate London, with smarting eyes and irritated lungs, was blinking, wheezing, and choking. Inanimate London was a sooty spectacle ...." Of course, the famously mood-enhancing London fog was actually the effect of high levels of coal-driven air pollution.

One expert has estimated the cost of this pollution. Roger Fouquet, a scholar of energy transitions, calculated the social costs (those not accounted for in market prices) of air pollution in Britain from the 1820s into the 1950s. In this period, coal consumption grew from 20 million tons a year in 1820 to 160 million in 1900. Air pollution deaths also grew: "higher concentrations generate proportionally more deaths."[1] In 1850, some 20,000 excess deaths (the difference between observed and expected) occurred due to air pollution. The figure grew to 50,000 by 1900, then fell to about 12,000 by 1950 as the government put emission controls in place.

Fouquet went further and attached a monetary value to the premature deaths. Using a metric known as the Value of a Statistical Life (VSL), he calculated the costs of air pollution to the British economy. VSL is a method for estimating the value people attach to the risk of premature deaths due to external factors like air pollution. In the 1820s, coal's early years, the damages amounted to about 4% of Britain's GDP; by 1850, after decades of growing coal use, it went up to 9%. The GDP share lost to pollution peaked in

1870–90 at some 15–20%. This meant that *up to one in every five dollars of income generated in Britain was lost to the health effects of air pollution.* In 1891, when pollution damages peaked, Britain's GDP was 100 billion pounds; the air pollution damages amounted to 17.5 billion.[2] From 1870 to 1890, air pollution alone made up two-thirds of the total costs (that is, the market costs plus the social or external costs) of coal. These figures do not account for emissions causing climate change, which almost nobody thought about at the time.

It was not until the Clean Air Act of 1956 that effective laws were adopted to deal with coal pollution in Britain, a century after it made health and economic sense. Bear in mind that Fouquet's study only estimates the health costs of air pollution, not other social costs: degraded land, contaminated water, damaged buildings and crops, harm from mercury and other toxics (the study only included particulates), and a century of emissions leading to climate change. Nor does it include coal miner deaths from accidents, which peaked in 1913 at over 1,700 – one in every 600 miners. In sum, over the time period studied, British coal production and combustion can be linked to some 4.5 million deaths.

Coal powered the industrial revolution and made prosperity possible, but at great cost. Britain was the birthplace of that revolution, but if we trace its effects over the rest of the world since the early 1800s, including the now rapidly growing economies in southeastern Asia and Africa, the tremendous costs of fossil fuel energy come into sharp focus. Other parts of the world now are dealing with the same problems Britain faced. And the pollution problems of developed countries are not over either.

The Climate, Health, and Economic Costs of Fossil Fuels

The scientific community paints an increasingly dark picture of a changing climate. The effects are felt in rising sea levels, increased flooding, extreme weather, prolonged droughts, new disease patterns, biodiversity and ecosystem harm, displaced communities, melting icecaps, crop failures, extreme heat, lost production ... the list goes on.[3] Worries about climate *change* have evolved into talk of a climate *crisis*. Climate fears cause high anxiety, especially among young people.[4]

Because the global energy system accounts for 74% of the emissions that accumulate in the atmosphere, it is the leading culprit. Climate is the clean energy catalyst, but it is not the only reason to decarbonize. The World Health Organization (WHO) puts premature death (mortality) and disease (morbidity) from air pollution near the top of its list of threats to human health.[5] In addition to its dire effects on the climate, carbon-based energy is a major cause of air pollution. Its harms do not fall equally. As is well documented, air pollution harms the poor and disadvantaged more than others.[6] Even in the US and Europe, which acted relatively early to control health-damaging emissions, the poor suffer more. In rapid-growth countries, as we will see later, the disparities often are worse.

Avoiding negatives like climate change and pollution is in itself a strong reason for achieving an energy transition by mid-century. Yet clean energy is about more than avoiding negatives. It is about realizing positives: the benefits of revitalized communities, life-supporting jobs, expanded knowledge, and breakthroughs to enhance well-being.[7] Some even see a potential for revitalizing democracy. But we will begin here with a discussion of the hidden costs of energy, what are known as social costs.

The True Costs of Fossil Fuels Through the Life Cycle

According to dictionary.com, something is *true* if it is "in accordance with the actual state or condition … real, genuine, authentic." If *hidden*, it is "concealed, obscure, covert." Both terms can be used to describe energy's social costs, as they arise from a changing climate, excessive water use, pollution, and community and ecosystem harm. It should be clear by now that the market price we pay for usable energy is not what it costs us. In paying the market price, consumers are paying for what it costs to find, produce, generate, process, and deliver energy to a point of use, plus taxes and other such items. A market price typically does not account for costs that are external to such activities, the *externalities*. Because they are not apparent in market prices, these externalities represent the *true*, *invisible*, or *hidden* costs of energy.[8] Because they affect society but are not accounted for in markets, they are termed *social* costs. Of course, even clean energy is not entirely clean. We live in a world of imperfect solutions, but some solutions are better than others. Clean energy describes technologies, sources, and practices that cause less harm and offer more opportunities than dirty energy. Clean energy has social costs, which will be reviewed in Chapter 5, but far fewer than those from fossil fuels.

Life-cycle analysis gives a picture of the non-market impacts of a product, technology, or process – from the initial design and sourcing of materials, to production, distribution, consumption, and disposal. Most external costs arising from fossil fuels come with their combustion. Fossil fuels do their job by driving turbines to generate electricity, delivering heat for industry, or sparking engines. Air pollution – health-damaging particulates, nitrogen oxides, and carbon dioxide, among others – are a main worry, but other damages not accounted for in market prices add to the costs.[9]

Table 2.1 lists the social costs of fossil fuels throughout their life cycle. There are four stages in this cycle: *extracting, transporting, burning,* and *disposing* of what is left. Extracting fossil fuels from where they have been stored for centuries itself harms health and the environment. Coal mining is a dangerous occupation, and conditions

Table 2.1: Summary of Social Costs of Fossil Fuels Throughout Their Life Cycle

Stage of the Life Cycle	Illustrations of Social Costs
Extracting	• Worker risks from accidents & disease • Ecosystem devastation; land disruption • Water consumption & contamination
Transporting	• Road accidents • Oil spills • Pipeline leaks
Burning	• Greenhouse gas emissions • Local air pollution: particulates, nitrogen dioxide, etc. • Local air pollution: toxics (mercury, chromium, etc.) • Water withdrawals for cooling
Disposing of Wastes	• Coal slurry/fly ash waste • Abandoned mines • Drinking water contamination

Source: Adapted from Union of Concerned Scientists, *The Hidden Costs of Fossil Fuels*, https://www.ucsusa.org/resources/hidden-costs-fossil-fuels.

for early miners were appalling. They improved because governments imposed stricter safety rules, but still the extraction process is risky. This goes beyond accidents. In the US, some 10,000 former miners were diagnosed with black lung disease between 1990 and 2000.[10] Risks in developing countries with lower health and safety standards are even higher.

Coal mining harms ecosystems. Especially devastating is mountaintop removal – blowing off the tops of mountains to get at coal seams. This was common in Appalachian areas of the US. It is hard to imagine a more ecologically destructive way of extracting coal. Mountaintop removal destroys landscapes, buries streams and valleys in waste, contaminates rivers and drinking water supplies, and coats areas in coal dust. Although the ecological harm is clear, health risks also are a concern. Blasting and processing coal releases hazardous, ultra-fine particulates made of silica and other materials that are linked to cardiovascular disease and to lung disease and birth defects.[11]

Producing natural gas through hydraulic fracturing (fracking) is controversial in the US, where it is practiced extensively and regulated unevenly. Although fracking has helped make the US the world's leading oil and gas producer and made cheap gas more available, it raises ecological and health concerns. Fracking involves injecting a complex soup of chemicals into the ground under high pressure; it uses lots of water, requires disposal of contaminated water, causes local air pollution and land disruption from drilling, and releases methane, a potent greenhouse gas.[12]

Because they are stocks rather than flows, fossil fuels must be moved from where they are produced and processed to where they are used. Coal and oil move on railroads, trucks, and tankers. Natural gas moves through pipelines or, over long distances, via ships in compressed and liquified form. At each stage there are risks of spills, explosions, and leaks. Major events of this kind include Mexico's Ixtoc 1 spill in 1979 (126–140 million gallons of oil); Uzbekistan's Mingbulak spill in 1992 (88 million gallons); and the *Amoco Cadiz* spill off Brittany in 1978 (69 million gallons).[13] Although at the production rather than transport stage, the BP Deepwater Horizon (2010) disaster ranks as one of the largest oil spills in history, when 134 million gallons spewed into the Gulf of Mexico over five months.

Still, fossil fuel combustion is where the most destructive impacts occur. The effects on climate change and air pollution are considered in the following sections of the chapter. For now, it is enough to say that fossil fuels are the primary source of carbon dioxide and health-damaging forms of air pollution. In addition, the carbon-based energy system is a major water user, especially in coal mining, oil and gas drilling, and in cooling thermal coal and nuclear plants that generate electricity. Energy competes with global agriculture in the amount of water used. As one commentator puts it: "In the US and Europe, more than half of the water drawn from nature is used for power generation."[14]

Finally, there are the waste by-products. Coal processing produces slurry wastes containing toxic metals like arsenic, mercury, chromium, and cadmium. Coal combustion leaves a by-product, fly ash, that must be stored or disposed of, often in lagoons that leak and pollute water. Earth Justice reports that, in the US alone, 110 tons of fly ash are produced annually and stored at over 1,400 sites, 70% of which are in low-income communities. Fly ash sites contaminate drinking water with toxins like arsenic and mercury, causing cardio-vascular, nervous, and lung diseases.[15] Like coal mining, oil and gas exploration and drilling activity damage vulnerable ecosystems and communities.

Over a decade ago, a Harvard Medical School study estimated the life-cycle costs of coal. It gave low and high estimates as well as a middle, best estimate. Its best estimate of annual costs in the US was $345 billion (2008 US dollars), with a low of $175 and high of $523 billion. When calculated on a kilowatt-hour (kWh) basis, this comes to 18 cents per kWh (9–24 cents). Most of the costs are from premature deaths due to air pollution ($187.5 billion) and from climate impacts ($61.7 billion). Other costs were $2.2 billion from land disruption causing carbon and methane releases; $5.5 billion from mercury health damages; $8.8 billion from abandoned mines; and $3.2–5.5 billion in public subsidies. Beyond these are the hard-to-monetize damages to ecosystems, land, species, and communities.[16] Per kilowatt-hour, the social costs of coal are much higher than its market price. In the absence of carbon pricing policies, these costs go unrecognized.

Directed by the Energy Policy Act of 2005, the US National Academy of Sciences (NAS) did its own study of energy's hidden costs.[17] It estimated the non-climate damages (health effects due to premature mortality) and those from illness (morbidity), including those from the 406 coal power plants then operating in the US, at

$62 billion, lower than the Harvard estimate. The figure is lower because the NAS used different assumptions about numbers of deaths from particulates and other pollutants. It estimated the health costs of the pollutants included in the study at 3.2 cents per kilowatt-hour, but did not include air toxics like mercury or many of the ecological effects included in the Harvard study, leading to lower estimates.[18] For CO2, estimates depended on the number used to calculate the **social cost of carbon**. Using the Obama administration's social cost of carbon measure of $50 per ton (discussed later), the NAS estimate was 5 cents/kWh. In 2020 dollars, these social costs amount to 11 cents per kWh, a few cents below the 2020 average US retail electricity cost – a significant tax on society.

Although exhaustive, the NAS study did not monetize all costs. Many toxics were not included on the health side, and some ecological costs were hard to estimate. Nor did it account for national security. For these reasons, the study "substantially underestimates the damages."[19]

The Climate Case for Clean Energy

Most people under the age of forty have grown up hearing about climate change. Although scientific concern about greenhouse gases and their effects goes back more than a century, the issue only established itself on the global policy agenda in the 1980s and early 1990s. *Policy agenda* is a term for describing issues that draw the attention of governments and others as requiring action. Many factors influence the policy agenda. In the case of climate change, it is shaped by the scientific evidence, by the increasing mobilization of global institutions and citizens, and by the impacts of flooding, droughts, extreme weather, heat waves, and sea-level rise.

The basics of climate change

The UN's Intergovernmental Panel on Climate Change (IPCC) defines it as "a change of climate which is attributed directly or indirectly to human activity that alters the composition of the global atmosphere and which is in addition to natural climate vulnerability observed over comparable time periods."[20] Climate varies over time; what matters is the external forcing effect of growing concentrations of greenhouse gases in the atmosphere that cause steady increases in average temperatures. Human activity has the forcing effect of causing abnormal rates of climate change. This has profound consequences. The causes are clear. The US Global Change Research program finds "it is extremely likely that human activities, especially emissions of greenhouse gases, are the dominant cause of the observed warming since the mid-twentieth century."[21] In its latest Assessment Report, the IPCC concludes that "carbon dioxide (CO_2) is the main driver of climate change, even as other greenhouse gases and air pollutants also affect the climate."[22]

Warmer temperatures cause all kinds of changes that disrupt natural systems, leading to the effects we hear about all the time. Warmer weather causes glaciers and mountain icepacks to melt, leading to sea-level rise and water shortages. Thermal expansion in oceans also causes rising sea levels. As the air holds more moisture at higher temperatures, the likelihood of extreme weather events grows, with a pattern of too much water at one time – flooding – and too little at another – drought. Ecosystems are thrown out of whack, causing species and biodiversity loss. With too little water, deserts expand and usable cropland shrinks. Food shortages increase as a result of less water and arable land, causing social and political instability and climate-change-induced migration.[23]

Climate change results from many forms of human activity, but fossil fuels are a major driver. They are the primary source of greenhouse gases, mainly CO_2. Other gases include methane from agriculture, landfills, biomass burning, and oil and gas production (16% of emissions); nitrous oxide (especially fertilizer and fuel combustion by-products) (6%); and fluorinated gases from refrigeration, industrial activity, and consumer products (2%). CO_2 from forestry and land use change (loss of tropical forests or expanding cropland) makes up 11% of greenhouse gases.[24] Table 2.2 lists the main greenhouse gases, their sources, and their varying global warming potential.

These gases accumulate in the atmosphere and are usually measured in parts per million (ppm): the number of CO_2 molecules for each million overall.[25] At the beginning of the industrial revolution, CO_2 concentrations in the atmosphere were about 280 ppm; by 2020 they had passed 410 ppm and were growing by 2–3 ppm each year. The global consensus is that such concentrations need to peak at 450–550 ppm (increasingly at the lower end) to avoid the worst impacts. The

Table 2.2: Main Greenhouse Gases and Their Global Warming Potential (GWP)

Gas	Main Sources	Percentage of Global Emissions	GWP
Carbon Dioxide	Fossil fuels & industrial processes	65%	(= the standard for stating the effects of other gases in terms of CO_2 equivalents)
	Forestry and land use	11%	
Methane	Agricultural activity Waste management Biomass burning Natural gas leaks	16%	28–32 times that of CO_2 over a 100-year period; 80 times that of CO_2 in the first two decades
Nitrous Oxide	Agriculture (e.g. fertilizer) Fossil fuels	6%	265–98 times that of CO_2 over a 100-year period
Fluorinated Gases	Refrigeration Industrial/consumer uses	2%	Thousands to tens of thousands of times that of CO_2, but over short time periods

Source: US Environmental Protection Agency (IPCC emissions data from 2010).

higher the concentrations, the more the heat-trapping effect, and the warmer the average temperatures. Think of it as an atmospheric bathtub: the flow of emissions builds up and causes warming. Yet this bathtub is not easily drained. CO_2 causes warming for centuries. The roots of climate change trace back to early coal use. Similarly, greenhouse gas emissions today will have impacts well into the future, making climate change truly an inter-generational issue.

Trends and emission projections

Historically, most emissions were from richer parts of the world: the US, Europe, Japan, and so on. More recently, growth in China, India, and elsewhere increased emissions in those countries, reinforcing a pattern in which economic growth relies on fossil fuels. As long as fossil fuels power economies and societies, this pattern will persist. Average temperatures are up 1.1 degrees Celsius since the mid-1800s;

at current trends, they may rise by over 2 degrees by 2100 if countries meet the commitments made as part of the Paris Climate Agreement (discussed in the next chapter), and by at least 3 degrees if they do not.[26] Figure 2.1 depicts the relationship between CO2e concentrations and average temperatures in the last 140 years.

The Global Carbon Project tracks emission trends closely.[27] In its November 2021 update, it reported that atmospheric levels of carbon dioxide equivalent gases were at 415 parts per million (ppm), some 50% above pre-industrial levels. Emissions fell in 2020 due to the global pandemic (an "unprecedented drop") but were rebounding. Total CO2 emissions (almost totally due to fossil fuels, with other gases counted separately) stood at nearly 35 gigatons, approaching four times the level of the 1950s. Of specifically CO2 emissions related to fossil fuels, 40% were from coal, 32% from oil, 21% from natural gas, 5% from cement, and the rest from gas flaring and other sources. To have a 50% chance of meeting the 2-degree target, the world cannot add more than 1,270 gigatons of CO2e to the global bathtub; to stay below 1.5 degrees, it cannot emit more than 420 gigatons; at current rates, that number will be exceeded in the 2030s.

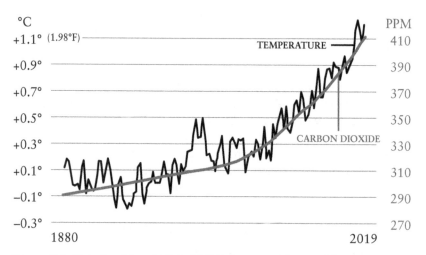

Figure 2.1: Relationship of Global CO2 Concentrations to Temperature
Note: Global temperature anomalies averaged and adjusted to early industrial baseline (1881–1910). Global annual average carbon dioxide. Source: Climate Central, https://www.climatecentral.org/gallery/graphics/global-temperatures-and-co2-concentrations-2020. Data sourced from NASA GISS, NOAA NCEI, ESRL.

Where is the world in cutting emissions? The 2020 *Emissions Gap Report* issued by the United Nations Environment Program found that the world is falling short of goals set at the 2015 Paris climate conference, even if all countries meet their current climate commitments.[28]

Impacts of a changing climate

Gina McCarthy, Environmental Protection Agency Administrator under Barack Obama and now President Biden's domestic climate coordinator, likes to stress that "climate action is about public health, not polar bears."[29] In 2020, University of Chicago energy economist Michael Greenstone testified before a US House of Representatives Committee on a study of the health impact of hotter weather. "Year after year," he testified, "heat records are broken all over the world."[30] His study estimates that, globally, warming may cause eighty-five excess deaths per 100,000 by 2100, and more in hot, poor areas.

This was a fine-grained analysis: the world was divided into 24,000 regions and mortality from hotter weather was assessed for each. The study accounted for the costs of premature deaths and identified the resources needed to avoid additional damages. For comparison, the estimated eighty-five excess deaths per 100,000 due to extreme heat is less than the global death rate for cancer (126 per 100,000), but higher than for infectious diseases (not including Covid-19) at seventy-four deaths per 100,000 and higher than diabetes (twenty-six deaths). Heat stress will significantly threaten public health as the climate changes.

Later I discuss the social cost of carbon, a method for calculating the monetary damages from greenhouse gases. During the Obama administration, the official US estimate for 2020 was a mid-point of about $50 per ton.[31] This accounted for a range of impacts, not just in health but in relation to sea-level rise, farm losses, lost productivity, and so on. Excess mortality costs were a fraction of that – some $2 per ton. In Greenstone's calculations, by contrast, mortality damages alone were *over* $36 per ton. "Put plainly," Greenstone concluded, "some of the most significant public health gains in human history could be achieved by cutting greenhouse gas emissions."[32]

More hot weather leading to heat waves is a top health and ecological concern. A few years ago, the environmental journalist

Jeff Goodell wrote that "Extreme heat is the most direct, tangible, and deadly consequence of our hellbent consumption of fossil fuels." The United Nations estimates that 1 million species globally are at risk from extreme heat and other climate impacts in coming decades. Parts of Asia, including India and Pakistan, may become too hot for human habitation. Goodell even predicts a "Hurricane Katrina of extreme heat" in which power failures and loss of cooling cause thousands of deaths from heat stress, possibly including in affluent areas like Phoenix, Arizona.[33]

Rising sea levels are also high on the list of concerns. Warmer temperatures cause sea levels to rise in two ways. First, melting glaciers and ice sheets add to the volume of ocean water. Second, thermal expansion causes water to expand as it warms. The US National Oceanic and Atmospheric Administration reported in 2021 that average global sea levels were 21–24 centimeters (8–9 inches) higher than in 1880. If the world cuts emissions, a best-case scenario is that sea levels will rise at least 0.3 meters (12 inches) by 2100; in the worst case they may rise by 2.5 meters (8.2 feet), an eventuality that "cannot be ruled out." In many US coastal areas, flooding occurs 300–600% more frequently than it did fifty years ago.[34]

These are just some of the impacts of climate change. In addition to the effects on public health, sea levels, and water resources are major hurricanes, typhoons, and other extreme weather events. New patterns of disease also emerge, especially as these effects combine with human intrusions into ecosystems. The oceans become more acidic. Species loss accelerates. And the longer we wait to act, the more severe the impacts and the higher the costs of adapting will be. Most worrisome are fears of *tipping points*, where the effects of climate change itself – warmer oceans that absorb less heat, or melting permafrost that releases methane – reinforce existing trends and accelerate the pace of change.[35]

Fossil Fuels and Air Pollution

Of the 7 million annual deaths from polluted air, the WHO calculates that 4.2 million are due to ambient (outdoor) pollution and the rest to household (indoor) exposure.[36] Both primarily result from the burning of fossil fuels. Outdoors, the pollutants of concern include particulate matter, especially fine and ultra-fine forms; nitrogen dioxide; sulfur oxides; ground-level ozone caused by nitrogen dioxide and volatile organic compounds; and air toxics like lead, mercury, arsenic, and chromium. A German study using satellite monitoring reached a higher estimate of 8.8 million deaths worldwide from air pollution, which subtracts nearly three years from average world life expectancy. The study concluded that "the majority of polluted air comes from fossil fuels."[37]

Household air pollution is an issue in poor and developing countries, especially in Asia and Africa. It is caused by indoor cooking stoves with fuels like coal, char, peat, and kerosene. It is a special threat to women and children, who spend more time indoors and suffer more exposure. To reduce the household threat, a priority is to shift to cleaner fuels like natural gas or expand electricity access, which is one of the UN's Sustainable Development Goals. Given global clean energy goals, natural gas is not a preferred solution. If possible, it is better to expand electricity access with renewables like wind and solar. Broadening electricity access simply by connecting villages to grids driven by coal power plants exchanges one source of health problems for another.

Children are "particularly at risk" from air pollution, given their smaller airways and faster breathing. In some parts of the world, children suffer from both indoor and outdoor pollution, in which case they "have little-to-no reprieve."[38] In poorer countries, like most of those in Africa, the health risk from indoor air pollution is falling while that from outdoor sources is increasing, underlining the importance of making energy more available but from cleaner sources, especially electricity.

Air pollution causes heart disease, strokes, lung cancer, acute respiratory infection, chronic obstructive pulmonary disorder, hypertension, and cognitive losses.[39] Fine and ultra-fine particulates are a special worry. Fine particulates are less than 10 microns in diameter (PM 10); ultra-fine are less than 2.5 microns, about 3% of the width of a human hair. The smaller the particle, the higher the risk, because small particles penetrate further into the body. Particulates top the list, but other pollutants are similarly damaging. Nitrogen dioxide combines with volatile organic compounds (VOCs) to form ozone, the second-most pervasive energy-related air pollutant. And nitrogen dioxide causes problems on its own.

One way to assess the benefit of clean air is in terms of its effect on life expectancy. The *Air Quality Annual Index* calculates the impacts of fine and ultra-fine particulate matter.[40] Pollution levels exceeding the WHO standard of 10 microns per cubic meter shorten life expectancy by an average two years. In areas with the worst air – Bangladesh, India, Nepal, and Pakistan – fine particulates subtract five years from life spans. In northern India, 250 million people lose an average of eight years of life expectancy.[41] Especially in South Asia and Central-West Africa, air pollution rivals or exceeds the harm from smoking, HIV/AIDS, poor sanitation, traffic accidents, and conflict and terrorism.

One study has calculated the premature deaths that could be avoided if the world eliminated fossil fuels. A decarbonized energy system would eliminate 41% of air pollution-related deaths. The US would avoid some 69% of air pollution deaths, Germany 61%, and China 55%. Numbers for those countries earlier in development are lower: 37% for India, 40% for Vietnam, 12% for Pakistan.[42] The percentages grow with more fossil fuel use. Globally, the study finds, two-thirds (65%) of air pollution deaths could be avoided with a decarbonized energy system. A similar study in the US found substantial health benefits from even low levels of renewable energy adoption. Benefits are highest in areas that rely mostly on fossil fuels for electricity. Many of these health benefits are lost when carbon capture and storage is used rather than renewable sources to mitigate carbon.[43]

Another study examined how meeting the 2-degree target set at Paris (discussed in the next chapter) without negative emissions strategies (like bioenergy with carbon capture and storage, a topic in Chapter 7) could improve health.[44] Focusing on particulates and ozone, it found that reducing CO_2 by an extra 180 gigatons – enough to meet the Paris goal without carbon removal – would deliver formidable benefits. Between 2020 and 2100, some 153 million deaths (plus/minus 43 million), 40% of them by 2060, could be avoided. The study documents local, near-term health benefits of decarbonizing, an antidote to the perception that climate change is a distant rather than immediate threat. It also underscores the importance of getting as much carbon as possible out of the global energy system. Politically, people respond better to issues affecting their near-term well-being. This study estimated the number of premature deaths that could be avoided in cities around the world, a way of making the data more personal. Table 2.3 presents the premature deaths that could be avoided by accelerating a transition to clean energy.

Table 2.3: Avoided Premature Deaths from Accelerated Reductions by 2100 (in thousands)

Moscow	320
Mexico City	290
Tianjin	290
Bangkok	260
Seoul	250
São Paulo	200
Los Angeles	130
New York	120
Tel Aviv-Yafo	93
Singapore	76
Johannesburg	67
London	53

Source: Drew Shindell et al., "Quantified, Localized Benefits of Accelerated Carbon Dioxide Emissions Reductions," *Nature Climate Change* 8 (2018), Supplementary Information, Table S1.

Also relevant here is a study from the Centre for Research on Energy and Clean Air, an Australian think-tank. It compiles results from a review of studies on the global health costs of air pollution related specifically to fossil fuels.[45] It estimates the total costs at 3.3% of global GDP, or $2.9 trillion. Most costs are due to avoided premature deaths ($2.4 trillion), but they also include costs of disability from chronic disease ($200 billion), sick leave ($100 billion), child deaths ($50 billion), and asthma and preterm births ($107 billion). This study calculates the GDP impacts of fossil-fuel-related air pollution. In many countries (China, Hungary, Ukraine, and India) these costs exceed 5% of GDP. The study also calculates health costs per capita, which are higher in richer countries – the US ($1,900 per capita), Germany ($1,700), South Korea ($1,100), and Japan ($1,000) – while the impact in terms of percentage of GDP is higher in the less affluent countries.

Disparate pollution exposures reinforce other global inequities. One review found that "air pollution is higher in poor communities."[46] Even in the wealthier countries, the American Lung Association finds, "Poorer people and some racial and ethnic groups are among those who often face higher exposure to pollutants and who may experience greater responses to such pollution."[47]

Such data and cost estimates are a standard way of looking at the damages from fossil-fuel-driven air pollution. Stories offer another picture. One source of stories is Beth Gardner's *Choked: Life and Breath in the Age of Air Pollution* (2019). From pollution emergencies in Delhi, to diesel emissions in London, to "air you can chew" from coal household furnaces in Poland, Gardner paints a picture of air pollution's effects. She describes a Krakow neighborhood as "a place that is, very literally, choking to death,"[48] and harm from traditional biomass in Malawi, where "pneumonia, often caused by cooking fires" from indoor charcoal and wood ovens, "is the single biggest killer of children under five."[49] Even in modern London, still reliant on diesel transport, the air is dangerous; 9,416 of city deaths were attributed to air pollution in a recent year.[50] Gardiner wants a "cleaner, healthier future ... in which breathing, life's most basic function, no longer carries a hidden danger."[51]

The Social Cost of Carbon

A useful way to think about energy is in terms of the *social cost of carbon* (SCC). This is a tool for estimating the economic harm resulting from a unit of CO2-equivalent emissions: "the present value of all future damages to the global society of one additional metric ton of carbon dioxide-equivalent greenhouse gases emitted today."[52] The SCC gives the cost of the damages as well as the positive benefits of not emitting CO2. In negative terms, it describes the monetary value of the harm that a ton of CO2-equivalent emissions causes into the future.

Calculating the SCC is no easy task. It requires an accounting of all the damages expected from each additional ton of carbon dioxide emitted in a year. This includes the costs of sea-level rise, extreme weather, human mortality and disease, flooding, droughts, farm losses, and more. Not only do we have to project what emissions will be, usually under different scenarios, we also have to attach a dollar value to those damages (*monetize* them) and convert them to current units. This occurs in four steps: predict future emissions, based on projected population and economic growth; model the impacts of those emissions in terms of health, ecological, and productivity damages; attach a monetary value to those impacts; and apply a discount rate to calculate the present value of damages occurring into the future, in terms of lost productivity, less growth, sea-level rise, heat stress, and so on.[53]

There is debate over just what the SCC should be. An interagency working group in the US estimated it at about $50/ton as of 2020.[54] The Trump administration recalculated this to a ridiculously low $1–6/ton.[55] Many economists think it should be higher than even $50 per ton, more like $200–300. For example, a 2018 study estimated a global SCC of a median $417 per ton, with a likely range (at a 60% probability) of $177–805. It also found that costs among countries and regions varied: India, the US, and Saudi Arabia suffered higher costs for each ton emitted globally.[56] The SCC provides a yardstick for evaluating the value of avoided emissions and a basis for taking climate action. The SCC numbers used in decision-making should be higher than they now are.

Analytical tools like the social cost of carbon are valuable for understanding climate change economically. The SCC does not tell us everything, to be sure. It does not tell us directly who suffers most, within or among countries or across generations. Nor does it convey a clear sense of the human costs of the disruptions experienced by communities, especially in poor countries. Not everything can be reduced to a number. But it does offer a tool for putting climate change damages on the economics table. Economics drive much of the decision-making in the world; if there is a case to be made by accounting for social costs, for the *hidden* costs of the global energy system, then it should be made.

From Burden Sharing to Opportunity Sharing

The growing and mostly irreversible effects of climate change offer good reasons to accelerate the energy transition. The benefits of decarbonizing, from better public health to smarter water use, ecosystem protection, and lower energy bills, strengthen the case. Clean energy avoids negatives but also defines opportunities – moving the world from *burden sharing* to *opportunity sharing*.[57] A look at two forms of opportunity sharing – good jobs and enhanced national security – will make the point.

Clean energy and jobs

Policymakers want to create economic and social conditions in which people have jobs. Unemployment has negative economic impacts, to be sure. But just as important are psychological well-being and a sense of belonging in society.[58] Of course, the form of employment matters. What is needed are jobs to provide a family-supporting income, financial security, and growth potential.

Major economic and social transitions – and clean energy is both – involve disruption and dislocation.[59] A case made against clean energy is that it means people will lose jobs. This is true to a degree: the shift away from fossil fuels to renewables, efficiency, and electrification displaces workers in the coal, oil/gas, transport, utility, and refining sectors. Because governments are accelerating the transition, many people (including myself) think that public policies should aim to ease this disruption and dislocation. Policies for offsetting economic and social disruption are discussed in Chapter 9.

One form of opportunity sharing is the creation of better-paying, durable jobs. This is the case Van Jones makes in *The Green-Collar Economy* (2008): We can "fight pollution and poverty at the same time" with a green-collar economy, creating "family-supporting, career-track" jobs that directly contribute to "preserving or enhancing environmental quality."[60] Clean energy delivers green jobs. Consider three studies making this case. All document that smart, clean energy investment creates more jobs per dollar than fossil fuel investment. A 2009 study found that six clean energy options – building retrofits, mass transit, smart grid, wind, solar, and biomass – yield two to three times as many jobs per dollar as coal, oil, or gas investment. Clean energy investments are labor-intensive, create more jobs, and rely on domestic sources more than imports. A 2014 update of this study reached similar findings.[61]

Another study found that "the renewable energy sector generates more jobs per megawatt of power installed, per unit of energy produced, and per dollar of investment, than the fossil fuel-based sector," and that "the bulk of studies ... conclude that a shift away from conventional energy sources could lead to substantial net employment gains."[62] Recent research confirms this. One million dollars put into renewables yields 7.49 full-time-equivalent jobs; a comparable investment in energy efficiency delivers 7.72 jobs. Investing the same amount in fossil fuels yields 2.65 jobs; "each $1 million shifted from brown to green energy will create a net increase of 5 jobs."[63]

IRENA studied the effects of business-as-usual relative to an aggressive clean energy scenario (*REmap*, discussed in Chapter 3) between 2018 and 2050. Overall employment increases in both. Energy efficiency accounts for most of the job gains in the first decade but for less after 2030. IRENA sees its clean energy strategy as increasing employment more broadly, not just in energy. By 2050, it forecasts a loss of 7.4 million fossil fuel jobs but a gain of 19 million jobs in renewables, efficiency, and grid modernization – a net gain of 11.6 million.[64] This is why job enhancement is cited as a justification for infrastructure proposals; clean energy investment also enhances the job growth that voters like.

In summary, the argument that clean energy is a job-killer is unfounded. The evidence is that efficiency, renewables, and a modernized grid alone will generate more jobs than are lost. Add in mass transit, new electricity transmission capacity, and hydrogen infrastructure and the job-killer case is even weaker. Clean energy is a path to green-collar, family-supporting jobs. This draws labor union support and makes clean energy ripe for public investment. The challenge is that politics in most places responds to the perceived gains and losses among specific groups, rather than to overall well-being.

Energy security

The origins of the clean energy transition go back to the oil crises of the 1970s, which highlighted many countries' reliance on imported oil. The oil embargos pushed France into nuclear, Denmark into wind, and led countries like Sweden to seek alternatives to fossil fuels. In the US, it stimulated creation of a Department of Energy and, for a time, a search for alternatives to imported oil. Even today, *energy autonomy* is a concern of governments seeking to reduce their dependence on fossil fuel imports. Russia's invasion of Ukraine in February 2022 brought a new urgency to worries about a dependence on foreign energy sources, especially in Europe, which depends heavily on imported natural gas and oil.

Dependence and vulnerability are hard to quantify, but dependence on fossil fuels imported from other countries introduces another category of costs. These mainly involve oil, which is traded on global markets. Oil is a global commodity. Although refining and distributing costs vary, crude oil is available at roughly similar prices in most of the world. Yet, as with other aspects of energy, imported oil has hidden costs. Although hard to calculate, they can be formidable.

Dependence on imports exposes countries to supply disruptions, price spikes, and other risks.

One effort to measure the military cost to the US of depending on global oil supplies comes from Securing America's Future Energy (SAFE), an organization of military and business leaders. In 2018, SAFE calculated that the US military spends "at a minimum" $81 billion each year protecting global oil supplies – 16% of the US defense budget. Spread over 20 million barrels of oil consumed daily in 2017, this amounts to over $11 per barrel, nearly 30 cents for each gallon of gasoline sold. In effect, the military budget subsidizes oil at a rate of nearly 30 cents per gallon. If the long-term costs of wars are included, this comes to 40 cents per gallon.[65] These are part of energy's social costs and, SAFE argues, should be included in determining energy efficiency standards and other policies. Countries may be able to reassess far-flung operations and reduce defense costs with a clean energy strategy.

These benefits extend beyond the standard economic and national security concerns. As well as reducing dependence on unpredictable supplies, clean energy supports a more resilient, reliable grid. Distributed energy combines residential, commercial, and local generation with more flexible distribution in microgrids and minigrids. Less centralized systems recover faster from natural disasters. Modernized electricity systems also aid resilience by enhancing cybersecurity.[66]

Renewable Energy Benefits: Measuring the Economics

A thorough account of clean energy's opportunity-sharing potential is the IRENA report *Measuring the Benefits*, which makes a green economy case: clean energy "can offer solutions for the dual objective of ensuring economic growth and the imperative to decarbonize economies across the globe."[67] It considers three scenarios: a *Reference case* based on current country plans; its own *REmap* scenario (in which the share of renewables doubles by 2030 from 2010 levels); and *REmap plus enhanced electrification* (with more electrified heating and transport).[68] This matters, since energy comprises some 6% of country GDPs, on average, and has profound economic and social impacts.

Challenging claims that clean energy hampers growth, the IRENA report finds that *REmap* would increase global GDP by 1% by 2030.[69]

The gains are, however, distributed unevenly. Japan, Australia, Brazil, Germany, and Mexico benefit most, followed by China, France, the US, India, and the UK. Major oil exporters (Saudi Arabia, Russia, Nigeria, and Venezuela) lose ground. While renewables grow, fossil fuels decline by 3%. This is why fossil fuel interests oppose decarbonization.

Of course, growth impacts matter less than the impacts on well-being. If we look at renewables through a sustainability lens – taking into account health, climate stability, energy access, quality jobs, and social equity, *including effects on future generations* – clean energy is a positive. As the IRENA report states, renewables "improve human welfare in a much broader manner and in a way that allows for future long-term growth and positive socio-economic development."[70] Among the positives:

- 11% fewer emissions by 2030 (nearly 16% under *REmap plus electrification*)
- A nearly 2% fall in volumes of materials used in the global economy
- Significant gains in public health due to lower air emissions, from outdoor sources but also from indoor ones like cookstoves in poor and developing countries
- A net gain of 6 million jobs and, under *REmap plus*, of 8 million
- Significant cuts in water use by the energy sector: wind and PV plants withdraw up to 200 times less water than conventional thermal coal and nuclear power plants
- Increased energy access in developing countries, with positive effects on development and well-being; some 60% of the new generation needed for universal access is likely to come from distributed energy: off-grid, stand-alone, and mini-grids

Can Clean Energy Lead to Fairer, More Equitable Societies?

Chapter 9 considers energy justice and democracy. Energy justice refers to the effects on economic and social equity. Energy democracy describes the distribution of political power and the ability of individuals and communities to manage energy choices. Many aspects of clean energy – lower social costs, decentralization, access, affordability – promote justice and enhance democracy.

Table 2.4: Clean Energy and Climate Action: Ten Benefits for Most
Countries

1 Avoid runaway costs of climate change	• Minimize effects & costs of extreme weather, wildfires, droughts, flooding, heat waves, sea-level rise, and others
2 Create jobs	• Clean energy investments generate more jobs • Efficiency & infrastructure are opportunities
3 Compete internationally	• Share in global clean energy economy (wind turbines, solar panels, multiple technologies) • Enjoy spillover effects of technology innovation
4 Improve public health	• Reduce local, health-damaging air pollution • Have fewer, less intense heat waves & wildfires • Avoid high temperatures that worsen air pollution
5 Save households and businesses money	• Save on heating, cooling, transport, manufacturing • Efficiency benefits of electric motors, green buildings, alternative transit
6 Enhance national and global security	• Reduce resource conflicts & climate migration • Reduce reliance on imported oil • Minimize climate & water-related political instability
7 Provide benefits to farmers	• Economic/resource benefits of sustainable farming • Income from carbon offsets, leasing land for wind & solar generation
8 Deliver benefits to low-income households	• Lower household energy bills • Lower transport costs, increased options • Lower adverse effects from air pollution
9 Preserve vital ecosystems & species	• Minimize damage from rising seas & extreme weather • Maintain essential ecosystem services • Reduce extinction rates due to climate change
10 Conserve water resources/clean water	• Reduce climate effects on resources • Minimize effects due to lost snowpack • Reduce water use for thermal power generation

Source: Adapted from Union of Concerned Scientists, "Top Ten Benefits of Climate Action," October 2, 2009, https://www.ucsusa.org/resources/top-10-benefits-climate-action.

An equity benefit of clean energy is the reduction of harm to vulnerable communities and a better quality of life. Low-income communities bear a large share of the air pollution burden. This is the case both in economically developed countries like the US and Europe and in developing countries, where pollution burdens are high at early growth stages. Chapter 4 considers the positives of energy efficiency for low-income households. Another benefit is the enhancing of prospects in areas lacking electricity access. Distributed electricity sources like solar PV allow countries to expand access when infrastructure is lacking. In sum, clean energy's advantages are compelling, nearly irrefutable. Table 2.4 gives a summary of the benefits in most any setting, but especially in developed countries.

Opponents continue to make their case: people are displaced by rapid change; renewables are costly; energy security is threatened by variable sources; investment patterns, infrastructure, and business models cannot handle a rapid pace of change. There is merit to such claims, but there are also responses. It all comes down to how we design and manage the energy transition, the subject of chapters 4–7.

Guide to Further Reading

Intergovernmental Panel on Climate Change, *Special Report: Global Warming of 1.5 Degrees C, Summary for Policy Makers*, October 2018.

Daniel J. Fiorino, *Can Democracy Handle Climate Change?*, Polity Press, 2018.

Todd A. Eisenstadt and Steven E. MacAvoy, *Climate Change, Science, and the Politics of Shared Sacrifice*, Oxford University Press, 2022.

National Academy of Sciences, *Hidden Costs of Energy: Unpriced Consequences of Energy Production and Use*, National Academies Press, 2010.

3
Getting the Carbon Out
Pathways to Decarbonization

The path to a decarbonized world is not an easy one. The fossil-fuel-based energy system was built up over the last 200 years. Carbon is locked securely into economies and societies. The energy transition requires the deployment of advanced technologies on a large scale, new business models and relationships, a redesign of complex systems, huge capital investments, and creative ways of financing them. It means that much of what is taken as given will have to change over the course of this century. Getting carbon out of the global energy system should lead to a cleaner, more efficient, more equitable world. In this way, decarbonization is a shorthand for mitigating the problems and realizing the opportunities set out in the previous chapter.

The question is how to get there and just what "there" means. One school of thought suggests that we should shrink the size of economies. This makes sense given the historically tight linkage of economic growth to fossil fuels. If growth is the driver, reversing it is a solution. Yet this line of thought, termed *degrowth*, has its problems.[1] Stabilizing and reducing the size of economies by, say, 20% should plausibly lead to at least a 20% cut in emissions. Given that decarbonization scenarios call for *at least* an 80% cut in emissions by 2050, with many calling for deeper cuts, this is too little. Although there are reasons to question the emphasis on economic growth, especially in affluent countries, deeper structural change is needed in order to decarbonize.

Three other problems with the degrowth option are worth mentioning. One is that no country has ever set out to, or is likely to, deliberately reduce the size of its economy. Some countries – Sweden and Denmark come to mind – are better at balancing growth with other social goals, such as economic equity. But their leaders do not claim to want to reverse growth. The second problem is that less growth means that less capital is available for investment. Third, as recent history has shown, degrowth adds to political polarization and makes collective action more challenging – and an energy transition requires collective action. Beyond the affluent countries, should the goal be to impoverish people?

The decarbonization scenarios examined here are either neutral or positive about growth. This partly reflects political calculation. Calling for what is perceived as a path to economic decline is not a winning political strategy. My analysis accepts that investment capital and political support are essential to a transition. The scenarios in this chapter reflect a *green economy* view: economies will expand, but "the composition and trajectory of growth must change through the adoption of new investments, policies, technologies, and behavior."[2] Even if growth rates fall or reverse, structural change is needed. A natural experiment came with the 2020 pandemic, when world CO_2 emissions fell by 5.4% and methane by roughly 10%.[3] This bought the world time. It delayed by a couple of years the time when emissions will exceed targets, but did not otherwise change our prospects.[4]

Decarbonization scenarios vary in three respects. One is timing. The targets generally focus on mid-century, although some have longer time frames. Another is how ambitious a goal should be. As greenhouse gases have built up, so have ambitions. Some envision a near total elimination of carbon-based fuels; others accept that some uses will remain, especially for the hard-to-decarbonize transport and industry sectors. A third variation is in the makeup of decarbonization. A fault line in energy debates is whether to aim for a system composed entirely of renewables or a system that also includes capture and storage (CCS, as well as bioenergy with carbon capture and storage, or BECCS) and nuclear. The issue here is whether or not renewables alone can meet global energy needs.

What is Decarbonization?

There are many reasons to decarbonize the global energy system, but worries about climate change are the main driver. For decades, governments have acted to protect people and the environment from air pollution. They have imposed technology controls that are authorized by laws like the US Clean Air Act or the General Ecology Law in Mexico. The rise of climate change on the policy agenda forced governments to design strategies for transforming the energy system itself. CO_2 cannot be reduced sufficiently with technology alone.

The principal body for bringing nations together on the issue is the Intergovernmental Panel on Climate Change (IPCC). Formed in 1988 by the United Nations Environment Program (UNEP) and the World Meteorological Organization (WMO), the IPCC assesses the scientific knowledge, evidence on vulnerability and impacts, and mitigation options for reducing emissions and enhancing carbon sinks. Its assessments play a central role in the Conferences of the Parties (COPs) held annually under the UN Framework Convention on Climate Change (UNFCCC), which was adopted in Rio de Janeiro in 1992. Since then, major agreements on climate have included the Kyoto Protocol (1997), the Copenhagen Accord (2009), and the Paris Agreement (2015), the last being most significant for current policy.[5]

The 26th Conference of the Parties, the most recent as of this writing, was held in Glasgow, Scotland, in 2021. Although receiving mixed reviews, it reinforced the strategy decided at Paris in 2015, and achieved agreements on financing, trading systems, and coal plant phase-out.

Article 2 of the UNFCC defines the *qualitative* goal of international climate policy as "stabilization of greenhouse gas concentrations in the atmosphere at a level that would prevent dangerous anthropogenic interference with the climate system."[6] This qualitative goal evolved into a quantitative one to limit the average global temperature increase to "well below" 2 degrees Celsius above pre-industrial levels, and "to pursue efforts to limit the temperature increase to 1.5 C." The consensus is that pre-industrial levels (early 1800s) were at about 280 parts per million (ppm) of CO_2 equivalent in the atmosphere; they crossed 400 ppm in 2014 and have been growing at a rate of about 2–4 ppm each year. The thinking is that a doubling of atmospheric concentrations above pre-industrial levels causes a 1 to 4 degree increase. Two degrees became a benchmark, with 450–550 ppm, and increasingly the lower number, seen as an upper limit on total concentrations. As the effects of climate change become more apparent, raising more alarm, the world aims to limit the temperature increase to 1.5 degrees Celsius to avoid the worst impacts.

The 2-degree target was endorsed at the Copenhagen COP in 2009 and adopted as a legal goal in the Paris Agreement. It is described accurately as "a political consensus based on scientific assessment."[7] As a result of IPCC assessments over many decades, the view is that exceeding the 2-degree target "would probably result in quicker and more unpredictable climate response, and even irreversible and disastrous consequences" for the world.[8] Economic analyses like the *Stern Review on the Economics of Climate Change* find that adaptation costs are more manageable if atmospheric concentrations are kept below about 500 ppm.[9] The 1.5-degree goal was added in Paris in an effort to remain below 2 degrees at the urging of less developed countries and small island states, many of which may cease to exist. Of course, the ideal would be to keep temperatures where they were in 2020, at about 1.1 degrees above pre-industrial levels. But given the levels of atmospheric greenhouse gases already built up, that train has left the station.

Decarbonization is thus the process of transforming the global energy system to reduce CO_2 emissions and limit the average temperature increase to 2 degrees, and ideally 1.5. This is a big deal. As discussed in Chapter 1, energy transitions historically were drawn-out affairs. This one must be compressed into decades. Just what has to happen is the subject of the scenarios reviewed in this chapter. The focus here is on staying at least to within 2 degrees, with attention to meeting 1.5 degrees. As the IPCC compares the two targets, "the transformation required to limit warming to 1.5 C are qualitatively similar but more pronounced and rapid over the next decades."[10] The differences are stark. At current emissions rates, the Global Carbon Project calculates, the world can emit 1,270 more gigatons of CO_2e before breaching 2 degrees Celsius; to avoid exceeding 1.5 degrees, that number is 420 gigatons – a limit that could be exceeded in the 2030s.[11]

Just how far does the world have to go in reducing emissions to avoid the worst effects of climate change and have a chance of remaining "well below" 2 degrees? A long way, although there are different ways of seeing this. The consensus is that the global economy should be largely or entirely decarbonized early in the second half of this century. On one view, this means aiming for zero-carbon emissions by somewhere between 2050 and 2080, which would require almost totally eliminating fossil fuels from the global energy system and expanding technological and natural sinks for carbon removal. Emissions should fall 45–50% by 2030 to have a chance of getting to net zero by mid-century.

Yet emissions do not have to fall all the way to zero. We can enhance our ability to remove carbon from waste streams and the atmosphere. Some methods of carbon removal are *natural*, like reforestation, afforestation, soil management, or ecosystem protection; others are *technological*, such as carbon capture and storage (especially linked with bioenergy) or direct air capture.[12] The problem is that most removal technologies are not yet proven as commercially viable. Given its focus on energy, this book concentrates on technologies linked to energy production and use, like CCS and BECCS. Still, natural removal methods are important. The Massachusetts Institute of Technology's Natural Climate Solutions Program argues that natural removal has the "potential to provide over one-third of carbon mitigation necessary for meeting a 2-degree warming target."[13]

Much decarbonization thinking reflects this view. It aims to balance emission cuts and removal to reach a goal of net-zero carbon, what one writer called in 2020 "the latest buzzword in the world of climate action."[14] Net zero is a state of *climate neutrality*, in which emissions get low enough for natural and technological removal methods to make up the difference, creating a climate-friendly equilibrium. The World Resources Institute calculates that to meet the 1.5-degree goal, carbon dioxide emissions should get to net zero by 2050–52, and all greenhouse gases must reach net zero by 2063–68. For 2 degrees, CO_2 should reach net zero between 2070 and 2085 and all greenhouse gases by 2100.[15] For our purposes, focusing on CO_2, the imperative is to reach net-zero carbon emissions between 2060 and 2070.

Of course, this is a goal for the *global* energy system. The developed countries (the OECD club) that have already gone through rapid growth phases are committed to net zero by 2050, while the still rapidly growing countries are aiming for a decade or two later. The EU, as noted in Chapter 1, has committed to 2050, as has the US under President Biden. Highly motivated countries are Finland (2035), Austria (2040), Germany, and Sweden (both 2045). China is committed to net zero by 2060, India by 2070. Even Russia and Saudi Arabia, two highly carbon-dependent economies, are aiming for 2060.[16] When and if they get there, these targets are a sign that the clean energy transition is indeed flourishing.

The 2015 Paris Agreement and later developments building upon it are the world's current best plan for decarbonizing. The cornerstone of the plan is that each participating country (191 in mid-2021) commits to a nationally determined contribution (NDC), in effect a national mitigation plan. The intent is that successive NDCs, submitted every five years to the UN, will be more demanding through a "ratcheting up of aggregate and individual ambition over time." This process of strengthening commitments will determine if the world achieves "a balance between anthropogenic emissions by sources and removals by the second half of this century."[17] Figure 3.1 depicts the levels of cuts in emissions that would be needed to meet 1.5-degree and 2-degree goals. The first trajectory is far steeper.

My focus here is on decarbonizing the global energy system to mitigate emissions and realize the other benefits outlined in Chapter 2. For this purpose, carbon removal is of interest only when linked to energy production and use, as in bioenergy with carbon capture and storage (Chapter 7). I consider how we can squeeze almost all the carbon out of the energy system by 2060 to 2070, reaching close

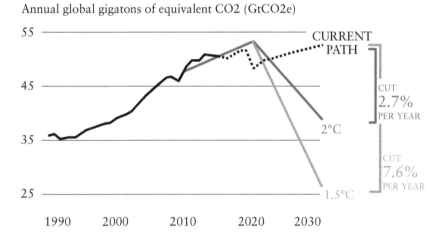

Figure 3.1: Levels of Emission Cuts Needed to Meet the Paris Agreement
Source: Climate Central, https://www.climatecentral.org/gallery/graphics/return-to-the-paris-agreement#modal. Data sourced from Climate Action Tracker, UN Environment Programme.

to but realistically not quite net zero. This "not quite zero" is a nod to reality and the law of diminishing returns, recognizing that some enhanced carbon removal will be necessary. On the practical side, there are parts of the energy system that are notoriously hard to decarbonize: heavy industry (cement, iron/steel) and some transport (aviation, shipping, freight). As for diminishing returns, the closer we get to zero, the more expensive it will be. The costs of going from 80% wind and solar electricity to 100% are higher than those of going from 60% to 80% (discussed in Chapter 6).

Getting nearly all the carbon out of the global energy system (industry, buildings, transport, and electricity), accompanied by natural and technological carbon removal, may get us to net zero. That would be a major accomplishment, perhaps the most impressive global cooperation in history. Considering just how that might happen from a big-picture perspective is the purpose of this chapter. The next four chapters then break decarbonization down into specific pillars of the clean energy transition. Table 3.2 at the end of this chapter gives a summary of these pillars.

Our big picture starts with how the world might totally decarbonize – an extreme form of *deep decarbonization*.[18] This scenario is daunting but worth examining in order to grasp the depth of the challenge. After that we consider a transition also based 100% on wind, water, and sun – with no nuclear, biomass, or CCS. This is controversial and involves many assumptions, but it offers a benchmark for other scenarios. The rest of the chapter reviews scenarios that use nuclear, biomass, and CCS to eliminate or offset remaining emissions. Since these scenarios vary, the chapter examines versions from the IRENA and IEA, as well as more specific ones for the US, Latin America, and India. The reviews differ, depending on the content of the analyses, but they set a foundation for the rest of the book.

I use *scenario* in the sense given by dictionary.com as "an imagined or projected sequence of events, especially any of several detailed plans or possibilities." Although some scenarios are more detailed than others, all set out "an imagined or projected sequence of events" for how a transition may unfold. Whether they are *imagined* or *projected* depends on how likely they are to occur. Scenarios may include specific *pathways*. Some analyses use the term "pathways" (e.g. the Deep Decarbonization Pathways Project) for their scenarios. I will use the term *scenarios* in all the cases.

Our look at decarbonization begins with the big picture: a triage of *easy*, *hard*, and *really hard* stuff. After that we consider 100% renewables followed by selected global and national scenarios.

The Easy, Harder, and Really Hard Parts of Decarbonizing

In *Our Renewable Future*, Richard Heinberg and David Fridley address the prospects for a global clean energy transition.[19] Their decarbonization goal is ambitious, as deep as we can go. They assess the prospects for zero-carbon energy by 2050. They reject nuclear as a "dead end" and see so many technical and economic problems with carbon removal that they eliminate that option too.

They propose a triage of *easy*, *harder*, and *really hard* stuff for decarbonizing. The easy, or at least easier, stuff starts with eliminating the use of fossil fuels to generate electricity. Coal and gas are replaced with wind, solar, and other renewables, and electrification is expanded to heat and cool buildings and power industry and transport. Vehicles charged with electricity generated by renewables are cleaner than conventional ones, and electric motors are more efficient (for reasons discussed later) than internal combustion engines (ICEs). The relatively easy stuff includes energy efficiency: retrofitting buildings, tightening building and appliance standards, and changing land use. Along with reducing energy use in the food system, Heinberg and Fridley see these changes as delivering another 40% cut in emissions over a few decades.

Then things get harder. Investments will have to be larger; complex systems must be modernized. As more electricity is generated by variable wind and solar rather than coal and natural gas, it is more difficult to integrate it into grids – to meet demand reliably and at reasonable cost. The more we rely on variable sources, the more challenging is the job of maintaining reliable power and balancing supply with demand. There are ways to deal with this: expand storage capacity, improve wind/sun forecasting, extend grids, manage demand better, increase efficiency. But once variable renewables exceed 60–80% of electricity generation, matching supply reliably with demand becomes a major challenge.

This hard stuff requires deeper change: "Electric cars aside, the transport sector will require longer-term and sometimes more expensive substitutions."[20] Among the options are replacing driving with alternatives and changing land use. Changes in commercial transport and industrial processes are necessary: electrified rail; fuel-cell trucks using green hydrogen; more closed-loop production processes; and eliminating fossil fuels in industry. This would cut emissions by another 40%. So doing all this on top of the (not so) easy stuff gets the world 80% of the way to decarbonizing.

Squeezing out the last 20% to reach zero-carbon energy is the *really hard* part of an energy transition. It will "take still more time, research, and investment – as well as much more behavioral adaptation."[21] Many industrial processes, such as cement production, require high heat levels that wind and solar cannot achieve without major technology advances or process redesigns. Changes in the food, communications, construction, iron/steel, and plastics industries are needed. Ocean shipping and aviation are also challenging. None of this is easy, but these are hardest to decarbonize.

This is a big-picture scenario for getting to a clean energy system based on zero fossil fuels. It sets the stage for the more detailed scenarios discussed in this chapter. Three points should be kept in mind: First, as we move closer to eliminating carbon, the difficulty and the costs increase. In economics terms, marginal costs increase as we approach zero. Second, zero carbon may or may not be our goal. Having residual fossil fuel uses may make sense, especially if carbon removal is feasible. Third, although the broad strokes of the energy transition are similar, the specifics "will look different in each country, so each country (and state) needs its own."[22]

As a big-picture look, *Our Renewable Future* envisions a complete overhaul of the global energy system, with no carbon-based fuels or nuclear energy by mid-century. This distinguishes it from the scenarios to follow, other than the 100% wind, water, and sun scenario discussed first. The others envision a continued use of fossil fuels, along with carbon capture and nuclear. In broad outline, however, they all reflect the big picture given here, with the distinction of easy, harder, and really hard stuff being precisely on target.

One point is worth stressing. Heinberg and Fridley doubt that total decarbonization can occur without adjustments in economic activity and lifestyles. They see the clean energy transition as having major implications for growth, finding that "whereas the cheap, abundant energy of fossil fuels enabled the development of a consumption-oriented growth economy, renewable energy will be unlikely to sustain such an economy."[23] The question is whether governments, economic interests, leaders, and voters will accept any such diminution of growth and lifestyles.

The next sections go deeper into three global scenarios: 100% wind, water, and sun; *REmap* (IRENA); and *Net Zero by 2050* (IEA). After that we look at more specific scenarios for Latin America, the US, and India.

A Technical Possibility? 100% Wind, Water, and Solar (WWS)

Our first scenario envisions a global energy system built entirely on renewables. It excludes fossil fuels, nuclear, and carbon capture. Nor is there use of biomass or biofuels. This scenario is the most aggressive among those reviewed here. This is not just pie in the sky, although some of the assumptions behind it can be challenged. The scenario is backed up by a detailed analysis for 139 countries.

Led by Mark Jacobson at Stanford University, the research envisions a 2050 world in which energy needs are met by 15% PV residential solar (percentages are rounded); 21% utility-scale PV; 10% concentrated solar; 12% commercial/government rooftop solar; 24% onshore and 14% offshore wind; 4% hydro; and geothermal, wave, and tidal energy amounting to just over 1%.[24]

The 100% roadmap relies heavily, as all the scenarios do, on reducing energy use with improved efficiency. Indeed, even with a shift to wind and solar, a decarbonized world is nearly impossible without huge efficiency gains. The 100% scenario cuts energy use by 42.5% below a business-as-usual projection for 2050. Over half of this is due to the "higher energy-to-work conversion efficiency" of electric heat pumps and motors. The rest comes from using less energy to meet the upstream demands of fossil fuel production (mining coal or transporting oil) and from improving overall energy efficiency.

In line with other roadmaps, 100% WWS delivers big gains for health, climate, and the environment as well as the economic benefits of lower costs and jobs. Full implementation by 2050 "avoids 1.5 C global warming and millions of deaths from air pollution annually; creates 24.3 million net new long-term, full-time jobs; reduces energy costs to society; reduces power requirements by 42.5%; reduces power disruption; and increases worldwide access to energy."[25] This could be done with available technologies, although many would have to be scaled up considerably. It is a clean energy advocate's dream. Figure 3.2 outlines the details and benefits of this scenario.

The 100% roadmap draws heavily on data, and models scenarios for twenty grid regions and 139 countries. The energy mix varies by region. Overall, the study predicts that the levelized cost of electricity

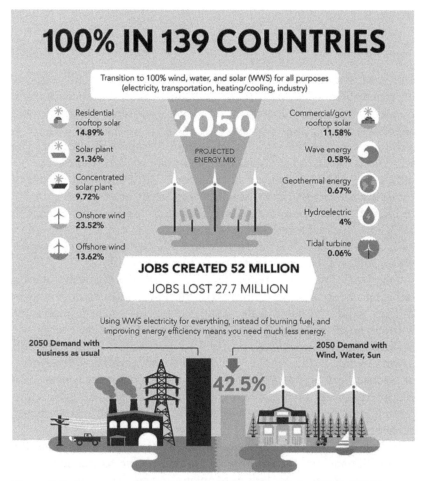

Figure 3.2: Summary of Energy Mix and Benefits of the 100% WWS Scenario

Source: Mark Z. Jacobson et al., "100% Clean and Renewable Wind, Water, and Sunlight All-Sector Energy Roadmaps for 139 Countries of the World," *Joule* 1 (2017), and The Solutions Project, https://thesolutionsproject.org.

per kWh is almost 1 cent less than in a business-as-usual projection (8.86 cents compared to 9.78). Even accounting for wind and solar electricity storage costs, the authors say, the LCOE would still be lower than under business as usual.

Whether or not renewables can expand on this scale is an issue. The scenario assumes we have fully electrified transport and industry, dramatically scaled-up technologies like green hydrogen, and rapid growth in affordable battery storage. But is an electrified system based on renewables feasible? The authors see this as a challenge, acknowledging that "the variability of WWS resources and associated costs of such uncertainty is the greatest barrier facing the large-scale, worldwide adoption of WWS power."[26] Still, they argue that "time-dependent supply can match demand at high penetrations of renewable energy without nuclear power, natural gas, or fossil fuels without carbon capture."[27]

The 100% scenario reviewed here was global; a similar one projected for the US drew a critical response. In *Proceedings of the National Academy of Sciences* (PNAS), twenty-one energy experts joined in a critique, claiming the study "used invalid modelling tools, contained modelling errors, and made implausible and inadequately supported assumptions." It added: "policy makers should treat with caution any visions of a rapid, reliable, and low-cost transition to entire energy systems that relies almost exclusively on wind, solar, and hydroelectric power."[28]

The 100% WWS scenario has drawn more than academic interest. Indeed, it became "a rallying cry for American progressives" after Jacobson co-authored an op-ed in *The Guardian* with Senator Bernie Sanders supporting a clean energy bill.[29] But critics saw this scenario as proposing a clean energy transition more disruptive and expensive than it needs to be, and as undermining the case for investing in needed technologies, including nuclear and carbon capture, that must play a role in decarbonizing.

The 100% scenario is useful in setting out a well-researched hypothetical for a totally decarbonized energy system. It provides a benchmark for evaluating other scenarios. We should welcome evidence-based additions to the debate, while recognizing the limitations in the analysis. Still, relying too much on any one scenario is risky. For this to succeed, everything has to work perfectly: investment at the right time and at needed levels; modernized grids and expanded transmission networks; the right balance of energy among transport, industry, and electricity; and smooth coordination of policy regionally and globally. Putting all of our eggs in a 100% renewables basket commits us to some options but forecloses lots of others. What if all of this does not happen?

There are economic and political issues as well. For reasons to be examined in Chapter 6, a 100% wind, water, and sun scenario would cost more than a more balanced, diverse approach. Overall, relying totally on variable renewables requires more installed electricity, given the lower capacity factors, than drawing on a diverse mix. A source generating electricity at 30–50% of its potential obviously has to have a larger capacity than one generating at 90%. Storage costs are higher; there has to be far more of it. It almost certainly costs more than a more balanced portfolio. Germany and Spain had to slow their energy transitions when costs rose rapidly. A balanced approach including nuclear energy and carbon removal may be part of the mix in places.

Aggressive But Possible? IRENA's REmap

The International Renewable Energy Agency (IRENA) prepares its own scenarios, which it calls *roadmaps*.[30] As a leading source of analysis on renewables, we might expect the IRENA to favor an energy system made up entirely of wind, solar, and other such sources. However, while the IRENA roadmap cuts fossil fuel emissions by over 70%, it does not eliminate them entirely. Make no mistake: it calls for dramatic cuts in fossil fuels by 2050. Oil falls by 77%, coal by 87%, and natural gas by 40% from recent levels. Per capita, CO_2 emissions fall drastically, from 4.6 tons per capita globally in 2018 to just over 1.0 by 2050.

The IRENA lays out two scenarios, a standard method in such exercises. The *Reference case* assumes all current and planned policies are carried out, including the nationally determined contributions (NDCs) under the Paris Agreement. In contrast, *REmap* defines a more aggressive plan that, says the IRENA, is technically and economically feasible, "based largely on renewable energy and energy efficiency, to generate a transformation of the global energy system that limits the rise in global temperature to well below 2 degrees Celsius above pre-industrial levels."[31] My focus being on decarbonizing, this section lays out *REmap* as a path to the Paris goals.

REmap starts with electricity, where renewables like wind, solar, and biomass are applied. Renewables would make up 86% of global electricity by 2050 and about two-thirds of final energy. After cleaning up electricity, the plan calls for a major scaling up of its use in transport, industry, and buildings, raising the proportion of total energy from electricity from 20% to 50% in 2050. Electrification "will be a key enabler to build a connected and digitalised economy and society."[32]

It all starts with cleaning up the electricity system. The renewables share in electricity more than triples, from 25% (in 2018) to 86% by 2050. Annual additions to global wind capacity grow over five-fold; those for solar PV more than triple. Passenger EVs would grow from a mere 6 million in 2018 to some 1.2 billion in 2050, and heat pumps increase more than fifteen-fold. Green hydrogen expands rapidly. Of the 86% of electricity from renewables, three-fifths is wind and PV, translating to a 2050 generating capacity of 6,000 gigawatts for wind and 8,500 for PV.

Energy efficiency is critical to *REmap*. Like other scenarios, the IRENA views the task as improving energy intensity – the energy needed to produce a unit of income. The rate of the intensity gains would have to go from about 2% annually in recent years to 3.2% by 2050.

Achieving these cuts requires major changes beyond increasing the electricity share of renewables. The share of total energy from modern, beneficial biomass more than triples from 5% to 16%, and use of traditional biomass (for indoor cooking and heating in poor areas) is eliminated. Energy from other renewables (geothermal, tidal, wave, concentrated solar) goes from less than 1 to 4%. Oil and coal fall to 13% and 3% of total energy. Total consumption falls, despite a larger global economy, due to better energy intensity. Figure 3.3 presents the difference between the *business-as-usual* scenario (the reference case) and *REmap*, describing the ways that the additional emissions cuts would be achieved as a result of the more ambitious strategy.

What investments would this require? As a comparison, for the *Reference case*, the IRENA calculates $95 trillion in investments through to 2050.[33] For *REmap*, to be consistent with the Paris goal, another $15 trillion is required. About one-third of this $110 trillion is for energy efficiency, one-fourth is for electricity infrastructure and renewables, and the rest goes to fossil fuels. The fact that efficiency

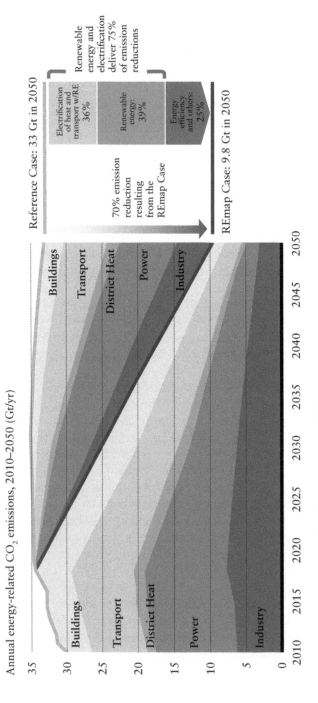

Figure 3.3: Comparison of *Reference Case* and *REmap*
Source: IRENA, *Global Energy Transformation: A Roadmap* (April 2019).

makes up most of the investment is testimony to its central role in decarbonizing.

The IRENA makes a strong case for clean energy's benefits. It estimates the avoided social costs at between $50 trillion and $142 trillion by 2050. Benefits fall into three categories: climate mitigation, valued at $17 to $46 trillion; better outdoor air quality, at $13 to $39 trillion; and improved indoor air quality, at $20 to $57 trillion.[34] This is in addition to the $15 billion saved by eliminating global fossil fuel subsidies, and it does not take into account many of the ecological damages of fossil fuels.

The International Energy Agency's *Net Zero by 2050*

Released in 2021, the IEA's *Net Zero by 2050* presents a "roadmap for the global energy sector." The IEA begins with the good news – that the number of countries (forty-four plus the EU) then having net-zero targets represented over half of global CO2 emissions. Even if they met all their commitments, however, that would still leave the world with 22 billion tons of emissions by 2050. And most existing targets are not backed by legislation or firm policy. Still, the IEA is cautiously optimistic on the prospects for net zero by 2050: there remains a "narrow but still achievable" path.[35] It notes that "All the technologies needed to achieve the necessary cuts in global emissions by 2030 already exist."[36]

The broad outline for decarbonizing is similar to *REmap*: aggressive efficiency gains; rapid wind and PV scale-up; accelerated EV growth (60% of new sales by 2030); substantial growth in electrified transport, buildings, and industry; and expansion in public EV charging stations from about a million in 2020 to 40 million by 2030. It calls for big cuts in fossil fuels, with coal making up only 1 percent of global energy by 2050 and oil and natural gas down by 75% and 55%. The IEA would ban new fossil fuel projects, aside from those already committed to, and fossil-fuel-based heating (to be replaced by heat pumps) by 2025. By 2050, renewables make up 90% of electricity, with wind and solar at 70%. To achieve this, PV has to grow twenty times and wind eleven above 2020 levels. Addressing a major worry, the IEA does not anticipate problems with critical materials like cobalt, copper, and manganese.[37]

Many people count on changed behavior for decarbonizing – less driving; more public transit, cycling, and walking; less electronics use; fewer transcontinental flights. However, the IEA report is realistic in viewing technology as making up over 95% of the clean energy transition. It expects behavioral change to account for just 3% of the needed emission cuts by 2030 and 4% by 2050.[38] It sees a pressing need for technology advances in battery storage, electrolysis for green hydrogen, and carbon removal. It calls for an 8% cut in absolute energy use by 2050, despite an expected doubling of the world economy and 2 billion more people to heat, cool, move around, educate, and care for. This will require $4 trillion by 2030. Figure 3.4 summarizes the four investment categories. The payoff is 9 million net new jobs and more rapid economic growth. This scenario calls for effective national action and "unprecedented international cooperation among governments."[39]

Selected Country Pathways

The transition has to be global, yet it will vary by country and region. Countries differ in their stage of development, economic composition, patterns of energy production and use, resource endowments,

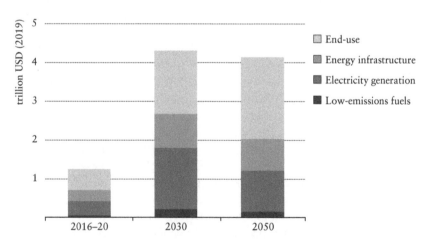

Figure 3.4: Clean Energy Investment in IEA's Net Zero Scenario
Source: IEA, *Net Zero by 2050: A Roadmap for the Global Energy Sector* (2021), p. 22.

and institutional and policy capacities.[40] Although the general paths are similar, the specifics vary. What works in Germany or France may not work in India, Kenya, or Brazil. I next consider three more specific scenarios for Latin America, the United States, and India.

Decarbonizing Latin America

The Deep Decarbonization Pathways Project (DDPP) was designed to demonstrate paths to clean energy in various countries.[41] In 2020, the DDPP released an analysis of pathways for six Latin American countries: Argentina, Colombia, Costa Rica, Ecuador, Mexico, and Peru.[42] DDPP country teams worked backward from the Paris goal of keeping the global average temperature "to well below 2C above pre-industrial levels" and pursuing "efforts to limit the temperature increase to 1.5C" by 2050–70. This translates into net-zero carbon by 2100 and to net-negative (that is, net carbon removal) later in the century. These countries start in different places. One measure of the difference is in terms of current emissions per capita. Peru and Costa Rica are lowest at 1.5 and 1.6 tons, respectively, the first due to its early stage of economic development, the second due to its having a clean electricity sector.

The scenarios for the six countries start by decarbonizing electricity with renewables (largely wind, solar, and hydro) and then electrifying transport, industry, and buildings. As a percentage of final energy, electricity rises from 28% to 82% by 2050. This allows overall carbon intensity to fall rapidly. Transport is always a challenge. Solutions include collective transport with better land use and urban planning, "efficient, affordable, safe and comfortable" public transport, and behavioral change, such as more telework. Collective and individual transport are more electrified. In this plan, the share of electrified transport ranges from a low 6% in Ecuador to a high 100% in Costa Rica; in between are Colombia (21%), Peru (43%), Argentina (53%), and Mexico (66%).[43]

A look at the decarbonized electricity mix reveals the variation among countries. Table 3.1 presents DDPP's plan to generate low-carbon electricity from eight options: wind, solar, hydro, geothermal, natural gas, bioenergy with carbon capture (BECC), and fossil fuels with capture and storage (FF+CCS). In Mexico, by 2050, nearly all electricity comes from wind and solar. Costa Rica, with one of the cleanest energy systems in the world, gets over half

Table 3.1: Electricity Mix (2050) in Five Latin American Countries in the DDPP

Country	Wind	Solar	Hydro	Geo	Gas	BECCS	FF+CCS	Nuclear
Mexico	36%	59%	–	–	4%	–	–	–
Costa Rica	29%	6%	52%	13%	–	–	–	–
Argentina	34%	13%	22%	–	–	–	–	31%
Colombia	5%	39%	35%	3%	–	3%	14%	–
Ecuador	6%	9%	44%	6%	–	30%	4%	–

Note: Peru not included. Geo = geothermal. – denotes negligible generation.
Source: Christopher Bataille et al., "Net-Zero Decarbonization Pathways in Latin America," *Energy Strategy Reviews* (2020), p. 7.

its electricity from hydro, a third from wind, and the rest from geothermal and solar. Argentina is the only country with nuclear power, which generates almost one-third of its electricity. Colombia relies on solar and hydro for electricity, augmented by fossil fuels with CCS. Ecuador uses BECC and hydro with the help of solar, wind, and geothermal sources.

As with other scenarios, efficiency is critical. Energy efficiency improves by 41–75% across the six countries in vehicles, residential and commercial buildings, and industry. What is left in 2050? The really tough sectors: road freight, aviation, and industries like iron and steel and cement. If there is a consistent theme in our scenarios, it is the challenge posed by heavy industry.

A Zero Carbon Action Plan (ZCAP) for the United States

In 2020, the Zero Carbon Consortium issued a plan for getting the US to net-zero carbon by mid-century.[44] It built on the Deep Decarbonization Pathways Project. This relies on the pillars of efficiency, decarbonized electricity, and electrifying transport and most of industry. The plan envisions using CCS in baseline power generation and to decarbonize any residual uses of natural gas in industry. Coal is gone within a decade. Petroleum use falls 90% and natural gas about 75% by 2050. Natural gas remains in "niche roles" as an industrial feedstock (an input to production), in transport, and for "a limited amount of … power generation needed to maintain reliability in an electricity system composed primarily of wind and solar generation."[45] The plan retains a modest nuclear capacity in order to generate baseload power and support the scale-up of wind and solar for electricity.

The plan breaks down uses for electricity, transport, industry, and buildings. For electricity, representing 32% of US energy-related emissions in 2019, it calls for wind and solar expansion, supplemented by other renewables, and nuclear generation. To maintain reliability, the plan maintains low levels of natural gas, but linked to capture and storage. Electricity grows from 20% to 50% of final energy.

Transport currently accounts for the most US emissions. In the Zero Carbon Action Plan, passenger fleets are battery electric, plug-in hybrid, or hydrogen fuel cell. Other transport – urban trucks and buses, rail, much of long-haul trucking, and some short-haul shipping and aviation – is electrified or converted to hydrogen. For long-haul aviation and ocean shipping, a low-carbon solution using advanced biofuels or synthetic liquids and gases produced with renewables is proposed.

Buildings and industry look different. Buildings could account for 12% of emissions through 2050, so new buildings offer opportunities for efficiency and electrified heating. All new buildings would use low-carbon materials and be highly efficient, as would retrofits. Industry accounted for 20% of energy-related emissions in 2020. Two-thirds of this was for energy, the rest for production feedstocks. Most light industry (food and textiles, for example) will be electrified in this scenario. For heavy industry, where current technologies do not provide electricity options and infrastructure is long-lived, CCS, green hydrogen, and synthetic fuels are likely solutions.

Where does this leave the US energy system in 2050? The economy's carbon intensity will have fallen by 95%, per capita energy use by 40%, and energy intensity by two-thirds – all signs of efficiency gains. Wind and solar will generate 90% of electricity, making up over 40% of the primary energy supply, with biomass at 12%. Coal, of course, is gone and oil nearly so. Natural gas falls from one-third of total energy to less than 10%, but is still used to balance grids, produce synthetic fuels, and meet hard-to-decarbonize heavy industry needs. Nuclear falls by about one-half (to about 10% of electricity). The plan foresees no expansion of geothermal and hydropower electricity generation.

Several features of ZCAP stand out: First, it maintains natural gas and nuclear. Natural gas helps cope with wind and solar variability and supports a cost-effective, reliable transition. Second, it uses carbon-neutral fuels to decarbonize heavy-duty, long-distance transport. Green hydrogen and synthetic fuels both play a part. Third, it relies on CCS to get carbon out of remaining natural gas uses. CCS rises from negligible levels to some 800 million tons per year by 2050.

To meet the ZCAP goals, US CO2 emissions have to fall at a relentless pace of 4% a year until 2050 – "a monumental transformation."[46] Still, in cost and technology terms, the scenario concludes that "carbon neutrality is an achievable outcome if the right policies are in place."[47] ZCAP underscores the complexity of the needed change, stressing that "successful implementation requires economy-wide coordination of the four foundational strategies across all sectors," including power generation, transport, buildings, and industry, as well as the critical role of government.[48]

The Zero Carbon Action Plan sets targets. In transport, half of all new light-duty vehicle sales should be electric by 2030, rising to 100% by 2040. Medium-duty electric sales are at 40% in 2030 and 80% by 2040; electric and hydrogen heavy-duty vehicle sales rise to 30% by 2030 and 60% by 2040. Electricity storage capacity reaches 20 GW by 2030 and 100 GW by 2040. Overall generating capacity doubles by 2050 to meet growing demand. Nuclear capacity remains the same, to provide baseload power, while natural gas capacity expands slightly to maintain grid reliability.

Deep decarbonization for India

Affluent countries like the US and EU matter in the transition, but the planet's fate depends even more on countries like India, where rapid growth, industrialization, rising living standards, and urbanization drive emissions growth.[49] Currently the third largest country source at 7% of world emissions, India could be at 13% by 2040.[50] It is early in its growth process and has an expanding population. Its economic growth rate from 1980 to 2014 averaged 6.2% a year. The good news is that electricity access in India doubled between 2000 and 2015, to 82%, and will reach 100% during the current decade.

A major challenge for India is the carbon intensity of its electricity sector. Coal generates 75% of electricity, and that sector accounts for 40% of emissions. "Coal currently is the backbone of India's energy systems."[51] The country's electricity system emits 721 grams of CO2e per kilowatt-hour, 50% above the world average.[52] Transmission and distribution losses are 23%, where the global average is 9%. Even with this heavy reliance on coal, per capita emissions are a fraction of those in the EU and US. With expected growth rates of between 5% and 10% in the coming decades, and population growth of some 300 million by mid-century, India's electricity system is a clean energy priority.

Clean energy progress in India is essential if the world is to meet its climate goals and if the air pollution burdens discussed in the preceding chapter are to be ameliorated. It all starts with the electricity sector, where the more ambitious of two deep decarbonization scenarios calls for a cut in carbon intensity of 60–85%. This can happen with more efficient transmission and distribution, and by replacing coal generation over time with renewables, but with substantial nuclear capacity to provide baseload electricity. After the electricity sector, at 40% of emissions, are industry at nearly 19%, agriculture at 14%, and transport at 13%. For industry, solutions lie with electricity, materials and energy efficiency, product substitutions, and linking fossil fuels to CCS. For transport, solutions lie in some electrification, freight system redesign, and intermodal shifts (from road freight to rail). Hydrogen could eventually play a role, although that would require a major scale-up. In terms of our decarbonization pillars, energy efficiency cuts across all sectors, renewables and nuclear clean up electricity, and much industry and transport shifts to electricity.

If CCS, CCUS, and BECCS (discussed in Chapter 7) are needed anywhere, it is in rapidly growing countries like India. With its carbon-intensive power sector and rapid growth in industry and construction (especially cement and iron/steel production), carbon capture technologies "could be an essential component of Indian energy policy." The International Institute for Sustainable Development (IDDRI) in Paris envisions using BECCS, described as "one of the most economical technologies" for capturing carbon, for reducing coal-related electricity emissions and for tough industry uses.[53] It sees an emission reduction potential of 800 million tons at a carbon price of $60/ton and a billion tons at $75/ton.[54] For this to happen, India requires strong research and technology development and policies for wind and solar, battery storage, green hydrogen, and grid integration/flexibility.

A recent study estimated that meeting one of the IRENA's aggressive clean energy scenarios in India will require, between 2019 and 2040, an investment of $5 trillion, most of it in low-carbon technologies (renewables, nuclear, EVs) and most of the remainder in efficiency and transmission/distribution.[55] The IDDRI estimates that, to achieve net zero by 2065, investment in India should be $120–140 billion annually from 2020 to 2045, for a total of $3.3 trillion. Policies must be "implemented in an integrated manner across ministries and sectors."[56] This highlights the need for strong policy leadership.

Countries like India are critical for decarbonizing. Its government has committed to net zero by 2070.[57] Whether this has a chance of occurring depends on national actions, financing, and technology developments around the world. Economic development is a priority for India and other developing countries. In equity terms, India's emissions per capita are well under half of those in the EU – a good performer among developed countries. Despite its very large population, India's cumulative historical emissions since 1850 are 5% of the total.[58] Its economy will grow in the coming decades, and wealthy countries have an interest in supporting its clean energy transition.

Lessons from the Decarbonization Scenarios

The first lesson to draw from these scenarios is that decarbonizing is difficult. All of them require dramatic change: huge expansions in wind, solar, and storage; transformed vehicle fleets; aggressive efficiency gains; and redesigned or new infrastructure. Describing what is possible is the point of the scenarios. But even the possible will not be easy.

A second lesson has to do with the outline of the transition. Coal disappears, quickly in developed countries but eventually in all of them. Natural gas may stay in the mix for a time, for industry use and to back up wind and solar in electricity. Electricity takes on a much larger role – in buildings, transport, and industry. As an energy carrier, not a primary source, electricity's role lies in applying the clean energy aspects of renewables. Among the latter, wind and solar do the heavy lifting.

Hydro has a role, especially in countries with resources and infrastructure. Biomass helps, although many forms are problematic. Geothermal adds to generating capacity where the right sources are available, as will ocean sources like wave and tidal at some point. Yet wind and solar are going to make or break the transition, with other renewables in complementary or niche roles.

The differences between scenarios and across countries and regions tend to focus on the role of nuclear, carbon capture, and technologies that are not yet commercialized. The 100% wind, water, and sun scenario appeals to clean energy fans, but it is risky. It could be hugely expensive, with rapidly rising costs that could stop the transition in its tracks. Both nuclear and carbon capture figure prominently in clean energy scenarios for countries like India. The prospects for nuclear, carbon capture and storage, and emerging technologies like green hydrogen are examined in Chapter 7.

A third lesson is that the pace of progress on energy efficiency has to accelerate. Efficiency could deliver from one-third to almost one-half of the needed decarbonization. To put it bluntly, if we are using much more energy by 2050, decarbonization is an elusive, perhaps unattainable goal. Efficiency's virtue, of course, is that *it saves money*. But it also takes money upfront, as investment that pays off later, so it is not a totally free lunch. This is the subject of Chapter 4.

Fourth, successful transitions have to be coordinated. The energy system has many moving parts and complex interdependencies. If connections among them are not coordinated, the energy system may get cleaner, but progress will fall short of our goals. Even fully electrified vehicles are not much help if they are charged with electricity from coal. Getting to electricity systems built on 80% wind and solar is unworkable if it is not facilitated by flexible grids and cost-effective storage. Technology investments in PV modules have to contend with the availability of critical minerals and technologies for reusing them at their end of life. In complex systems, all the parts are interrelated.

Another lesson is that many of the assumptions and projections in the scenarios rely on technologies that are unproven at a commercial scale. This is clearest with nuclear and carbon capture and storage. Assuming that carbon removal technologies can deal with any overshoot in emissions on a large scale is a leap of faith. Of course, the same may be said of advanced nuclear, beneficial biomass, offshore wind, and others. Still, we need to squeeze carbon out of the energy system, not only as a hedge against carbon removal costs but to reap the climate, health, and other benefits.

What is also clear is that government must play a big role. To be sure, the private sector is critical for investment, technology innovation, infrastructure development, and more. Can the private sector lead, with government in a supporting role? Probably not. This is the first major energy transition driven by social costs – by the climate, health, and ecological damages of the current system. Government has to put a price on these costs because markets do not. Just as governments had to protect air and water quality, they must now push markets to account for the social costs of fossil fuels.

Furthermore, much of the transition involves public goods like basic research.[59] We call these *public goods* because they are hard to commercialize, and private firms cannot make money from them. Only government is in a position to invest in such areas as basic materials science and information technology. Governments around the world are already deeply involved. Virtually all aspects of the energy system are affected by government policies on subsidies and taxes, research funding, grid operation and connection standards, the construction and operation of infrastructure (roads, grids, and pipelines), and by multiple other policies, from regulatory standards to carbon taxes. In many places, governments own and operate major utilities and energy firms.

Finally, government has to drive this transition because there is so much at stake. Change on this scale is not just a matter of investments, markets, incentives, and technology. Working as it should, government leads collective action based on a vision of the future. Energy is essential in the modern world; how it is produced and used affects health, economic well-being, social equity, and security. Only government can lead collective decision-making in determining the content of the transition, its pace, its allocations of costs and benefits, its investment priorities, and the parts citizens and institutions will play.

The next four chapters consider decarbonization in more detail. Common to all scenarios are the clean energy pillars of energy *efficiency*, a major ramping up of *renewables* in electricity, and *electrifying* as much as possible of transport, industry, and buildings. Each is examined in the next three chapters. Most scenarios include roles for nuclear and carbon capture. These options, and that of using green hydrogen for transport, industry, and storage, are discussed in Chapter 7. Previewing the next four chapters, Table 3.2 introduces the clean energy pillars and their complementary technologies.

Table 3.2: Summary of Decarbonization Pillars in the Next Four Chapters

Decarbonization Pillar	What Must Happen	Why It Should Happen
Pursue aggressive energy efficiency	• Limit or stabilize demand growth • Focus on end-use efficiency; transmission & distribution • Reduce energy & carbon intensity	• Higher demand undermines the transition • Efficiency is low-cost carbon mitigation • Efficiency saves money
Ramp up renewables	• Accelerate scale-up • Focus on wind & solar as workhorses • Use other renewables for special needs & when available • Lower overall costs	• Far cleaner for electricity • Reduces climate, health, ecological impacts • Realize economies of scale
Electrify what you can; then seek zero/ low-carbon options	• Harness clean energy from renewables • Use distributed energy • Enhance grid integration & flexibility	• Key for transport, industry, & buildings • Adds to efficiency (electric motors) • Manages toughest sectors
Develop and evaluate complementary technologies	• Technology research & development • Define appropriate roles • Adopt policies for scale-up when suitable	• Support a transition • For baseload and specialized needs • Carbon removal is likely necessary • Green hydrogen has multiple roles

Guide to Further Reading

Richard Heinberg and David Fridley, *Our Renewable Future: Laying the Path for One Hundred Percent Clean Energy*, Island Press, 2016.

Mark S. Jacobson et al., "100% Clean and Renewable Wind, Water, and Sunlight: All Sector Energy Roadmaps for 139 Countries of the World," *Joule* 1 (2017), 108–21.

Scott Victor Valentine, Marilyn A. Brown, and Benjamin K. Sovacool, *Empowering the Great Energy Transition: Policy for a Low-Carbon Future*, Columbia University Press, 2019.

Deep Decarbonization Pathways Initiative, *Climate Ambition Beyond Emission Numbers: Taking Stock of Progress by Looking Inside Countries and Sectors* (2021).

4

The Invisible Resource
Energy Efficiency

The scenarios reviewed in Chapter 3 share many elements. They all call for a transition from a fossil-fuel-based energy system to one built on renewables. Although some scenarios envision a role for natural gas, even coal for a time, both sources are linked to carbon capture, and coal falls to or nearly to zero. Because renewables generate electricity – and eventually may produce green hydrogen at scale – all scenarios call for a scaling up of electricity for transport, industry, and building needs.

They share another element: more efficient energy use. If energy use keeps growing some 2% a year, as in recent decades, our goal of getting the carbon out is all but lost. Without dramatic efficiency gains, mid-century energy use could be 50% higher than in 2020.[1] The more energy the world uses, the greater the demands on variable sources like wind and solar. Electricity grids will be stretched, potentially undermining their functioning and increasing costs. Everyone pays more, not just because they use more energy, obviously, but because the costs of ensuring reliable electricity systems will be higher.

The International Energy Agency finds that efficiency "could provide more than 40% of the abatement required by 2040 to be in line with the Paris agreement."[2] The 100% wind, water, and sun scenario incorporates a 40% cut in energy use. The American Council for an Energy Efficient Economy (ACEEE) thinks the US could cut energy use in half by 2050.[3] The US Zero Carbon Action Plan calls for a 40% per capita improvement in energy efficiency by 2050.[4] IRENA's *REmap* relies heavily on efficiency. Energy efficiency is possibly the most overlooked yet most critical clean energy pillar.

The potential of energy efficiency is not currently being realized. In its *International Energy Efficiency Scorecard*, the ACEEE notes that efficiency "remains massively underutilized globally despite its proven benefits and its potential to become the single largest resource for meeting growing energy demand worldwide."[5] Opportunities not only to decarbonize but to save money are being overlooked, leading to depictions of energy efficiency as the *invisible resource*.

This chapter's themes are, first, that aggressive energy efficiency gains are essential. Any major consumption growth dooms the enterprise. Second, that efficiency is one of the least costly and most effective ways to decarbonize. Third, that it saves money while delivering other benefits: cleaner air, healthier ecosystems, less climate impact, affordability, and jobs. Most of the time, efficiency is the cheapest, fastest way to get the clean energy benefits reviewed in Chapter 2.

The chapter considers the outlook for energy demand and use in the coming decades, discusses the role of government and policy, and offers examples of top energy efficiency performance in specific countries. Before getting to those topics, however, it is worth considering some energy efficiency basics to get the lay of the land.

Energy Efficiency 101: What We Need to Know

To begin with we can distinguish energy *conservation* from energy *efficiency*. The US Energy Information Administration defines conservation as "any behavior that results in the use of less energy."[6] Turning off lights, using mass transit, limiting air travel, living in small houses, walking to work – all illustrate conservation. In contrast, efficiency means "using technology that requires less energy to perform the same function." Replacing old windows, trading internal combustion vehicles for electric ones, installing smart meters in offices, or building homes with adequate insulation are all steps to improve efficiency. Conservation plays a major role in cutting energy use, especially if we can change behavior. Given my concern with a clean energy system, my focus is on efficiency. This is not to say that conservation is unimportant and should not be pursued, but it is not the issue here.

To a limited extent, both conservation and efficiency are within individual control. Someone may decide to stop driving to work and take a bus or the subway. But what if there are few public transit options, as in American cities like Houston and Phoenix? I may want to do my part for clean energy by purchasing an electric vehicle, but I am not doing much if the electricity it runs on comes from coal. It costs money to replace insulation or windows, install smart metering, buy an EV (which, in 2022, still cost more up-front than standard cars), or invest in high-efficiency heating and cooling. Given these *system constraints*, the ability of individuals to conserve energy or to use it efficiently is limited.

We may also distinguish activity, structural, and efficiency effects.[7] *Activity effects* are changes in behavior – driving more, using less cooling – and economic output, like more manufacturing. *Structural effects* are changes in a category of activity, such as the number of appliances or vehicles or the amount of building space in use. Both activity and structural effects increase energy use as an economy grows. *Efficiency effects* describe the amount of energy used per unit of activity. People may drive the same number of miles, own the same number of appliances, or produce the same quantity of goods, but use less energy, mainly due to technology improvements. As growth occurs, activity and structural effects lead to more energy use that has to be offset with efficiency. Both activity and structural effects are essential in limiting energy use, but the focus here is on the efficiency effects.

A standard measure of energy efficiency in an economy or in parts of it is *energy intensity*, which is the energy used to deliver a unit of GDP. Another way of looking at this is in terms of *energy productivity* – the economic output generated from a unit of energy. These are inversely related: as intensity falls, productivity improves. Between 1990 and 2015, the EIA reports, world energy intensity fell 32%, a good thing. It declined by 28% in OECD countries and by 40% in non-OECD ones.[8] Generally, energy intensity and productivity are better in the mature economies, as noted in Chapter 1. Indeed, in 2015, OECD countries used 12% less energy per dollar of economic output than non-OECD ones. Early industrialization is energy intensive, as the US and Great Britain demonstrated in the last century. Still, early growth is when the options for improving energy intensity are at their best.

Economies tend to become more service-based than manufacturing-based over time, which helps reduce their energy intensity. A country's economic composition (its makeup of goods and services produced) affects energy use, as do factors like weather and population density. Energy intensity is usually higher in areas with large temperature variations (requiring more heating and cooling) and long distances between urban areas. Policies also matter. OECD economies tend to have strict standards for buildings and appliances, which reduces intensity and improves productivity. The more energy-intensive economies include Russia and Canada, partly as a result of their cold climates. Still, Russia stands out as one of the least energy-efficient countries in the world, largely due to weak policies.[9]

Figure 4.1 shows changes in energy productivity from 1990–2015. Aside from Brazil and the Middle East, all parts of the world improved. The country with the lowest energy productivity in 1990, China, showed the most gain: 133%. This reflects changes in the composition of its economy – less manufacturing and more services – but technology also played a role. The next largest gains were in non-OECD Europe and Eurasia, where the emergence of countries from the Soviet bloc led to modernization and enhanced efficiency. Despite their high starting point, OECD countries in Europe also

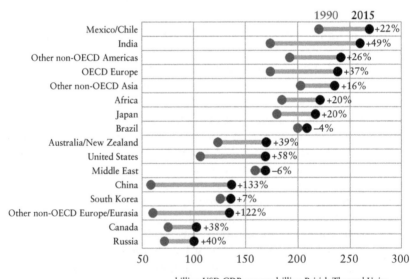

Figure 4.1: Energy Productivity in Selected Countries, 1990–2015
Source: EIA, "Today in Energy," July 12, 2016, https://www.eia.gov/todayinenergy/detail.php?id=27032.

gained by 37%. While this tells a positive story about energy productivity, the challenge is to reduce *absolute* energy use. Despite steady gains in productivity and intensity, economic and population growth increased energy use by some 50% from 1990 to 2015.[10] The world has to decouple growth from energy use.

The world has been and will continue to be more efficient in using energy. The steady gains in energy intensity are evidence of that. Yet these gains are more than offset by the activity and structural effects of economic growth. In 2018, the IEA reported that energy efficiency in most countries had improved by 15% since 2000, while consumption had increased due to higher activity (more cars and trucks, larger homes, more manufacturing). Activity effects accounted for 5% of the growth, structural effects for a much smaller 0.3%. Activity effects in particular offset the efficiency gains. Yet improved efficiency helped. Without that, the IEA estimated, the one-third growth in energy use between 2000 and 2017 would have been two-thirds, and CO_2 emissions would have been 15% higher.[11]

The term *efficiency gap* is useful, referring to "the difference between the cost-minimizing level of energy efficiency and the level ... that is actually realized."[12] This gap represents the money-saving opportunities that are not realized. For reasons discussed later, people often lack the information or the motivation they need make cost-effective investments in energy efficiency.[13] Public policy can help overcome these barriers with information, incentives, training, or by facilitating access to the up-front capital needed for investments. These policy options are reviewed later in this chapter and in Chapter 8.

Offsetting many efficiency gains are **rebound effects**. When people use less energy to do what they want or need to do, they can afford to use more of it. Having fuel-efficient cars frees us, in terms of cost, to drive more miles; energy-efficient refrigerators make owning a second one more affordable. Extend this across an economy and the losses add up. Jesse Jenkins examined nearly 100 peer-reviewed studies in affluent countries and found that rebound effects had eroded 20% to 70% of the savings from efficiency measures.[14] In emerging economies, where most economic growth will occur, the rebound effects could be significant.

From the studies of rebound effects, Jenkins draws the positive conclusion that, although we lose some efficiency gains, we improve energy productivity, which "simply means we are getting more energy services out of our resources than ever before."[15] This is good news in developing countries, where enhanced energy access and efficiency enable improved well-being. In developed countries, it highlights a need for conservation and the use of such complementary tools as access to financing or carbon pricing to offset the rebound effects.

The International Energy Outlook

The Energy Information Association is the statistical arm of the US Department of Energy (DOE). DOE was created in the wake of the 1970s oil crises to promote effective, coherent US energy policies. The 1970s saw the birth of the modern energy efficiency movement, because the oil shocks caused prices to spike and raised fears of energy insecurity. These shocks were the first global disruption in the status quo that effectively punctuated the energy policy equilibrium. The second, of course, beginning a decade later, was concern about the impacts of climate change.

A regular EIA publication is its *International Energy Outlook*, the purpose of which is to estimate trends in coming decades.[16] Energy use is driven by economic growth. The EIA's *Reference case*, discussed in Chapter 1, projects a global economic growth rate of 2–3.7% annually from 2020 to 2050, most likely about 2.8%. This varies across countries, however, and especially between the OECD and the emerging or still-poor non-OECD countries. A growth rate of 5% in India, for example, will drive steadily higher energy use.[17]

The main energy end uses are for industry, transport, and buildings. Industry includes manufacturing, construction, refining, mining, and agriculture – the *really hard stuff* from Chapter 3. Transport includes road passenger and freight, rail, aviation, and ocean shipping. In its 2019 *International Energy Outlook*, the EIA estimated growth for each of these between 2018 and 2050. Industry energy use will grow by over 30% to make up half of the total.[18] Building energy use in non-OECD countries will grow at five times that of the OECD to make up more than one-fifth of the total.[19] In sum, growth means more manufacturing, construction, resource use, mobility, and urbanization. Most of this growth will be concentrated in non-OECD countries. For example, transport energy use is expected to remain flat in the OECD but to rise 77% in non-OECD countries.[20]

This does not mean that OECD countries do not have to use energy more efficiently. Their economic and energy growth rates are lower, but they are energy intensive overall, especially in the US and Canada, which consume relatively more energy per capita than most parts of the world.

What may be possible

The EIA's *Outlook* is a reference case of what is likely to occur if something does not change significantly. The challenge is to change what will occur from a likely to a far more favorable outcome. The contributions efficiency can make to this change are evident in the IEA's *Efficient World Scenario* (EWS), issued in 2018. The IEA assumes that by 2040 world GDP will double from 2017 levels, population will be up by 20%, and there will be 50% more building space. Absent efficiency gains, decarbonizing in this case would be difficult. Yet existing cost-effective technologies could enable a 12% emissions cut by 2040, "40% of the abatement required to be in line with the Paris targets."[21]

The emerging economies, where the bulk of the growth that drives energy use will occur, would be 50% less energy intensive under the EWS. China and India alone would reap one-third of the savings. In terms of the main categories of energy end uses, the EWS means that:

- Although transport activity will double due to growth, energy demand would be flat by having more stringent fuel economy standards and far more efficient vehicles.
- Technical innovation in industry (electric heat pumps; efficient motors), combined with energy management systems and information technology, may deliver savings, especially in industries like food, beverages, and textiles. The value of production for each unit of energy used could double.
- The world has 60% more building space, a result of growth, but without additional energy demand. Standards for new buildings, retrofits for existing ones, and appliance standards make residential and commercial spaces 28% and 37% less energy intensive, respectively.

How would this occur? Through smart, effective policies as outlined later in this chapter and in Chapter 8. A start will be removing all energy subsidies, which offset its costs and reduce incentives to efficient use. *The Economist* magazine estimated in 2015 that worldwide fossil fuel subsidies from 1980 to 2010 led to extra energy use that accounted for 36% of global carbon emissions in that period. It added that "Recent research has shown that [subsidies] enrich middlemen, depress economic output, and help the rich, who use lots of energy, more than the poor."[22]

The ACEEE did a US efficiency analysis in 2017. Titled *Halfway There*, its theme is in the subtitle: *Energy Efficiency Can Cut Energy Use and Greenhouse Gas Emissions in Half by 2050.* Big greenhouse gas reductions due to efficiency are possible in light- and heavy-duty vehicles (an emissions cut of 30%), freight transport (25%), aviation (50%), and new homes and buildings (70%). Other reductions are realized in appliances and equipment (13%), upgraded existing homes and buildings (30%), and enhanced industrial efficiency (30%).[23] To achieve this, the ACEEE strategy uses stringent building codes; grid technology investments; strict fuel economy standards; a shift to electric vehicles; appliance standards, product labeling and benchmarking; and behavioral incentives. It goes beyond technology with options like public transit, walkable neighborhoods, biking, and ridesharing.

Policies matter, but higher levels of investment are needed. The global EWS calls for over $500 billion in annual efficiency investments in the early 2020s, up from half that amount in 2017, to over a trillion dollars from 2026 to 2040. By following the advice in *Halfway There*, the ACEEE estimates, the US alone would realize $700 billion in efficiency savings by 2050.

Is degrowth the answer?

Given that economic growth is driving energy use, why not just limit it? Without a doubt, smaller economies and slower growth would reduce energy use. If the global economy were to remain the size in 2050 as it is now instead of possibly doubling, it would use less energy. Energy use in what the ecological economist Herman Daly calls a "steady-state economy" would be far less than it is now.[24] There even is a school of thought calling for economic *degrowth* to reduce not only energy use but other aspects of growth – materials use, transport, lifestyles – that stress the planet. Degrowth writers want to shrink economies in affluent countries, stabilize growth in emerging ones, and make room for poor countries in Africa and southeastern Asia to achieve to a better quality of life through growth.[25]

No doubt affluent societies could use less energy by adopting less competitive and consumptive lifestyles – smaller houses, less air travel, better materials use, and smarter transport. Research on happiness finds it is associated with income, but only up to a point. The research also tells us that wealth alone does not define well-being; the level of inequality in wealth and incomes also accounts for a variety of social ills.[26] Indeed, happiness and well-being do not rise inexorably with income once certain needs are being met.[27]

Nevertheless, to hang the transition to clean energy on stopping or reversing economic growth is risky. Wisely or not, growth is highly valued. It is seen as a measure of national well-being, status, and power. People depend on growth to create jobs, fund retirements, educate children, and provide opportunities for the next generation. Modest growth creates investment capital for financing an energy transition. It provides revenues to conduct research, fund technology innovation, modernize grids, expand public transport, and conduct other needed activities. Moreover, as noted in the first chapter, simply having smaller economies does not accomplish much. Clean energy requires major investments, policy leadership, new business models, and much more. Reducing the size of the global and national economies will not on its own deliver the transformation the world needs.

Still, citizens and governments should reflect on what a good quality of life means and on the challenge of reconciling lifestyles with planetary constraints, including energy production and use. Having less energy-intensive lifestyles and consuming less energy are steps in the right direction. But this alone does not get us to where we need to be.

Shouldn't Efficiency Be a No-Brainer?

There is an old joke about two economists – one young and one old – walking down the street one day. The young economist looks down and sees a $20 bill on the street and says, "Hey, look a 20-dollar bill!" Without even looking, the wiser colleague replies, "Nonsense. If there had been a 20-dollar lying on the street, someone would have already picked it up by now."[28]

Probably anyone who has taken an introductory economics class has heard this story. Most of us would likely make the effort to bend over and pick up the bill. Still, there are reasons why someone would not. For people of ample means, 20 dollars may be too small a sum to bother. Someone may be distracted by other things on the street, lack the physical ability to pick up the bill, or notice a bigger bill nearby. They may feel obligated to leave the bill for someone else to find.

The same holds in energy. At any level of decision-making – in a private firm, a household, or an office – things get in the way of making rational choices. Information on energy-saving options may be unavailable. People may not appreciate the financial benefits of saving even small amounts of money. Other investments may promise higher returns, or other problems may distract people from the mundane task of getting more output from each unit of energy. People get stuck in routines or lack expertise. They are not always rational in taking advantage of an opportunity or they simply have other priorities.

Energy efficiency, as was noted above, is described as an invisible resource. People can see solar panels or wind turbines and marvel at how technology enables a smart electricity grid. Efficiency usually does not offer such opportunities. It is hard to appreciate what we do not use: "there are not many ribbon-cutting or groundbreaking ceremonies because individual investments in energy efficiency tend to be both small and invisible."[29] As David Goldstein observes, efficiency "lacks a trade association that can communicate its value consistently and repeatedly."[30]

Self-interest is a motivator, but it is not the only influence on decision-making. People can be motivated to act based on values. Yet the invisibility of efficiency extends to social behavior as well. Many people who care about climate change and advocate for renewable energy and electric vehicles may be less attuned to how efficiency cleans up the energy system. Making people aware of the potential contribution of energy efficiency to the clean energy transition helps speed it along.

Why do people not take advantage of money-saving opportunities? They do to a degree, or energy productivity would not improve. But they miss opportunities as well, even economically rational ones. Indeed, experts give attention to the *barriers* to energy efficiency. These barriers are "tenacious, interconnected, and deeply embedded in social fabrics, in institutional norms, in regulations and tax codes, and in modes of production."[31] It is easier to identify them than to overcome them. Still, only by identifying them can we devise policies for using energy more efficiently.

Many barriers are financial. Efficiency may make economic sense, but it requires up-front investment. High-efficiency products (state-of-the-art heating and cooling systems, or low-emissivity windows) require capital, which may be in short supply. Even when the payback period on the investment is short – three years or less – efficiency competes against other investments and may not be a priority. Investing in advanced technology involves risk and uncertainties. Policies for addressing the up-front costs, by shifting them from capital to operating costs, can help reduce the efficiency gap. They are part of the energy efficiency toolkit.

Information barriers also matter. Most of us are not experts on energy efficiency technologies, services, or products at the household level. Even within industrial firms, expertise lies more in a knowledge of production processes and technologies than in energy efficiency strategies. Outside experts on efficiency, in turn, may lack industry-specific knowledge. Another factor is *information asymmetries*, in which one side in a transaction (perhaps a building owner or salesperson) has more information about a product or technology than the one purchasing it. Policies that overcome this lack of information – product labels, training, audits – also are part of the efficiency toolbox.

Another barrier is *principal–agent disconnects*, which occur when those making decisions about products or upgrades are separate from those bearing the burden of high energy bills. This matters in commercial buildings, where space often is rented, but it relates to households as well. Renters, for instance, who have not had a hand in designing buildings or selecting appliances, lack the incentive or authority to make upgrades, even when they are paying the bills.

Government policy may undermine efficiency. Energy subsidies that keep prices below market levels discourage conservation and efficiency. If burdens are a concern, there are tools other than subsidies. Government regulations that made sense at one time can undermine efficiency later on. In the US, for example, policies regulating public utilities encouraged them to sell electricity and expand supply infrastructure to justify charging higher rates; many states have now recognized this as a problem, *decoupling* sales from prices and requiring utilities to promote efficiency.[32]

Research in behavioral economics expands our knowledge of the efficiency gap. People are prone to ignore information or to find it confusing, to stick with the status quo as creatures of habit, to be more concerned about avoiding losses than earning gains, and to put limited value on benefits accruing well into the future – all of which discourages efficiency.[33] A role for public policy is thus to create incentives and conditions that will overcome such barriers and stress society's interest in efficiency. Energy efficiency is a private good – saving it reduces costs – but it is also a public good that lowers demand and smooths the path to a transition.

A 2012 survey by the Economist Intelligence Unit, of over 400 executives in commercial, residential, and industry real estate and construction, sheds light on these barriers. The most-cited barriers to investment in existing buildings were a lack of demand for energy-efficient buildings (37%) and the perception that efficiency does not enhance building value (28%). One-fourth of those surveyed cited a lack of know-how, short investment horizons, and problems in securing credit.[34]

This chapter focuses on end-use efficiency – how consumers meet their needs but use less energy. But there also are ways of increasing production efficiency. One method is highlighted in Box 4.1, on combined heat and power.

Box 4.1: Combined Heat and Power (CHP)

CHP (cogeneration) uses waste heat from electricity generation to increase power plant efficiency. Most plants convert only about 40% of fuel to electricity, which means they are wasting up to 60% (the US DOE gives an average 34%). CHP systems recover the waste heat to create steam, hot water, or even cool water to deliver energy. CHP is especially effective in combination with district heating and cooling systems that serve sites with several buildings (like a hospital; see Box 6.2). With district heating, CHP can achieve fuel efficiency rates of 70–85% or higher.[35] It cuts costs and emissions (carbon and local pollutants) with economies of scale and by using existing infrastructure. Combined heat and power is a way to squeeze more electricity and heat production out of a generation source, and thus to cut emissions.

The Costs of Energy Efficiency

Costs matter for two reasons. First, efficiency ranks among the cheapest ways of reducing emissions and avoiding energy's adverse health and ecological impacts. Indeed, many measures save money in simple market terms, even before social costs are counted. Second, some measures are more cost-effective than others: efficiency can be gained more or less efficiently. The top choices are actions or investments that save money directly by cutting energy costs. After that, it makes sense to look at options for realizing other goals, such as less pollution or more employment.

The following sections examine two ways of comparing efficiency to other decarbonization measures in cost terms. The first looks at what it costs to mitigate one ton of carbon dioxide equivalent. The second uses the concept of the levelized cost of electricity introduced in Chapter 1.

Energy efficiency and the costs of carbon abatement

Over a decade ago, the consulting firm McKinsey compared a range of options for cutting carbon emissions. Its *cost of abatement analysis*, presented as Figure 4.2, organized information on the costs of various mitigation options – efficiency measures like LED lighting and state-of-the art appliances, wind and PV generation, reforestation, and CCS. The vertical bars in the diagram estimate the costs of mitigating (that is, not emitting) one ton of CO2. Options below the line have negative costs: they save money. A bar's width represents its mitigation potential. The costs of avoiding a ton of CO2-equivalent (tCO2e) emissions are given per euro.[36]

Energy efficiency dominates the left side of the diagram: LED lighting, efficient buildings and appliances, building insulation. In the mid-range of mitigation costs are measures to increase supply with non-carbon sources: wind, solar, and nuclear. The more expensive right side includes CCS. Much has changed in the last fifteen years, of course. Many options cost more, with inflation. But a 2017 update found that the costs of some options fell more than anticipated, among them PV, wind, battery storage, and LED lighting. The costs of others, such as CCS, fell less than expected.[37]

A more recent study stresses the distinctions between *static* and *dynamic* mitigation costs.[38] Static costs are a snapshot in time.

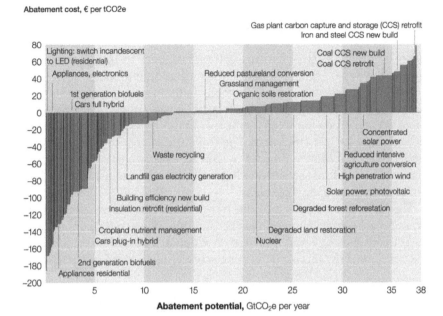

Figure 4.2: Carbon Cost of Abatement Curve for Mitigation Options
Note: The curve presents an estimate of the maximum potential of all technical greenhouse gas (GHG) abatement measures below €80 per tCo2e if each lever was pursued aggressively. It is not a forecast of what role different abatement measures and technologies will play.
Source: McKinsey, "A Revolutionary Tool for Cutting Emissions, Ten Years On," April 21, 2017, https://www.mckinsey.com/about-us/new-at-mckinsey-blog/a-revolutionary-tool-for-cutting-emissions-ten-years-on.

Dynamic costs account for changes over time: economies of scale that lower costs per unit; effects of locking in new, more efficient technologies; research and development spillovers, where knowledge gained on one technology or practice applies to others; and network externalities, when progress in one area has positive effects in others. For example, fast-charging stations promote electric vehicles, and less costly utility-scale batteries help integrate PV into electricity grids. Dynamic costs provide a fuller comparison of mitigation options. Over time, mitigation costs fall as economies of scale, technology learning, and spillover effects occur. It is worth noting that one reason for the rapid global decline in the cost of solar panels was Germany's national strategy, the *Energiewende*. Feed-in tariffs and other policies stimulated PV scale-up in Germany, which then "subsidized lower cost solar for the rest of the world."[39]

The levelized costs of energy efficiency

The cost per ton of CO2 abatement is one way to compare options. Another is the levelized cost of electricity. Recall that LCOE is the lifetime market cost of a kilowatt-hour of electricity. For electricity, onshore and offshore wind costs fell in the last decade, and those for PV dramatically. LCOE also applies to efficiency: each kWh not used is one that does not have to be generated. Consumers save money, to be sure, but everyone is better off. Less generation means less infrastructure, fewer peaks in electricity demand, and fewer social costs.

In their 2015 study *Green Savings*, Marilyn Brown and Yu Wang found that 57% of regulatory options for efficiency save money.[40] Several fall into the no-brainer category: incentives to support combined heat and power (2 cents per kWh); electric motor standards for industry (3 cents); residential building codes (1 cent); and commercial building codes (3–5 cents). In 2014, the ACEEE calculated that the LCOE for efficiency measures was lower than all of the available options for expanding energy supplies, presented in Figure 4.3.[41]

If our goal is meeting energy needs in a clean, low-cost way, efficiency is the place to start. Figure 4.3 presents data the ACEEE collected from the forty-eight largest investor-owned utilities in the US. It records the levelized cost of electricity of efficiency relative to supply options in megawatt hours, based on 2018 data. Efficiency stands out as a clear bargain. Efficiency's LCOE compares very favorably with wind, solar PV, and natural gas combined cycle generation, and is much cheaper than the other generation sources. The efficiency LCOE includes the costs of low-income support programs, making it an even more favorable comparison. Even with declining wind and PV costs, energy efficiency is a low-cost way to meet energy demand, not to mention the other benefits of reducing energy use through improved efficiency, such as less pollution, more reliable grids, and lower energy burdens for households.

This does not mean we should avoid wind and PV. They will replace retiring fossil fuel sources. It does mean we should get what we can through efficiency, then turn to cleaner sources. Without enhanced efficiency, the prospect is that renewables will make up

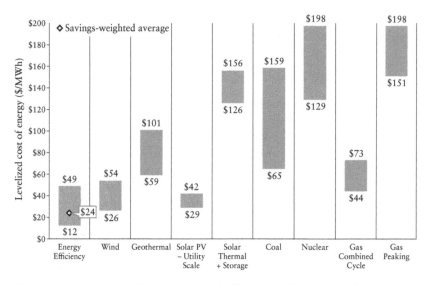

Figure 4.3: Costs Per Kilowatt-hour of Efficiency Relative to Supply Options

Source: Charlotte Cohn, "The Cost of Saving Electricity for the Largest U.S. Utilities: Ratepayer Funded Efficiency Programs in 2018," ACEEE, June 2021, https://www.aceee.org/sites/default/files/pdfs/cost_of_saving_electricity_final_6–22–21.pdf.

for higher energy use but will not replace fossil fuels; this delays the transition and makes it more costly.

Bear in mind that all of this applies only to the market costs of efficiency. The social benefits (avoided climate, health, and ecological costs) make efficiency an even better bargain. Add in the economic and job benefits, and the case is overwhelming. There are indeed a lot of 20-dollar bills lying on the street. The question is: How do we get more energy efficiency?

Why Government Policy Is Important

As with the other elements of the energy transition, government must play a central role. If we accept that one purpose of government is to promote citizen well-being, ample benefits may be gained.[42] An obvious one is lower costs and more efficient economies. Energy expenditures are estimated at 6–8% of world GDP, although this varies with fluctuating energy prices, especially for oil.[43] In the US, energy spending in recent years was about 6% of GDP. Energy is second only to health care as a spending category in most economies. Lower energy costs free up resources and investment for other social goals – health care, education, poverty alleviation, and of course clean energy. Especially in carbon-based systems, each megawatt of electricity or barrel of oil not used also means there is less climate damage, fewer pollution deaths, and less ecological degradation.

Efficiency also makes the transition away from fossil fuels easier and cheaper. Even if the world only uses 50% more energy in 2050, it will be much harder to eliminate coal, reduce natural gas to special roles, and capture the remaining carbon. Look at it this way: if energy use remains stable between the early 2020s and 2060, a seven-fold increase in wind and solar capacity would increase their electricity share to over 80%; if energy use grows by half, that same amount of wind and solar amounts to only 53% of generation. That is a big loss in renewable capacities.

Steady efficiency gains improve the ability to integrate wind and solar into grids.[44] It lowers the peaks in daily loads that require expensive natural gas peakers, which provide extra generating capacity at peak demand. To the extent that efficiency lowers and smooths out daily and seasonal demand, it makes variable renewables more feasible.

Low-income households in particular benefit. Anything that cuts emissions, of course, helps the disadvantaged households and communities that bear disproportionate shares of pollution and climate burdens. Yet efficiency promotes social equity in another way – by lowering energy costs. Low-income groups spend more of their income on energy: heating and cooling, appliances, getting to work. A 2016 study of US cities found that low-income households spent 7.2% of earnings on energy, while high-income ones spent 2.3%. Energy burdens were three times higher among low-income groups, in some cases up to 20% of income.[45] Burdens may be higher in developing countries with more inequality and less electricity access, if there is access at all.

Low-income households also are less likely to be able to reap efficiency's benefits. They lack the resources to invest in high-efficiency products: advanced heating and cooling systems, low-emittance windows, electric vehicles. Efficiency is a classic illustration of the old maxim of needing money in order to save it, and low-income households often don't have it. Related to this, larger proportions of low-income households rent rather than own residences. They depend on building owners or managers, who lack a stake in cutting energy costs, especially if tenants pay the utility bills. Having fewer resources, low-income groups tend to drive older, less efficient cars. Since for them energy is a financial burden, anything that cuts costs is good for social and economic equity.

Finally, energy efficiency should be a centerpiece of *green economy* strategies at any level of government.[46] A green economy links energy, health, social, environmental, and economic goals in complementary, ideally even synergistic, ways. Few issues present opportunities similar to those of energy efficiency. The health and environmental value is clear. Another benefit is stimulating employment: efficiency creates jobs. Making a case for governments to invest in efficiency in pandemic recovery programs in 2020, the IEA stressed that investing 1 million dollars in energy efficiency generates an average of up to fifteen jobs.[47] Grid modernization, building efficiency, and urban transport have similar job impacts. And these opportunities can be mobilized quickly.

Public policy is essential for progress. Later, I review policies that are helpful in promoting energy efficiency. First, it is worth looking at examples of leading countries and best practices.

Which Countries are Leading on Energy Efficiency?

Countries vary in their commitment to efficiency. Every other year, the ACEEE evaluates the twenty-five largest energy-consuming economies.[48] It uses thirty-five policy and performance indicators in four categories – buildings, industry, transport, and national effort – to measure high-level activities like energy intensity, the strength of national goals, and energy-efficiency spending.

About 40% of the indicators focus on performance, like energy intensity, the other 60% on policies. Ideally, better policies lead to better performance, but such factors as population density and climate also affect performance. To illustrate, a policy indicator assesses vehicle fuel economy standards; a performance indicator compares actual efficiency. Assessing both policy and performance provides a fuller assessment. Policy indicators are government actions; performance indicators are the outcomes. Table 4.1 lists selected indicators.

Germany and Italy tied for top spot in the ACEEE's 2018 *International Energy Efficiency Scorecard*, followed by France, the United Kingdom, and Japan. Countries that are (or recently were, in the case of the UK) part of the EU did well. Table 4.2 presents the top-ranked countries in the four performance categories. Germany led in national effort, Spain in building efficiency, Japan in industrial efficiency, and France in transport. Italy placed in the top three in all but one category, giving it a high overall ranking. Along with Germany and France, Italy led in the policy rankings. Taiwan, China, and Italy were the top countries in performance.

Table 4.1: Selected Indicators from the ACEEE *International Energy Efficiency Scorecard*

National Effort
   Change in energy intensity
   Energy efficiency spending
   Efficiency at thermal power plants
   Tax incentives and loans

Buildings
   Residential and commercial building codes
   Appliance and equipment standards
   Building retrofit policies
   Energy intensity of commercial and residential buildings

Industry
   Energy intensity of industry
   Mandatory energy audits
   Combined heat and power installed capacity
   Standards for industrial motors

Transport
   Passenger vehicle energy efficiency
   Fuel economy standards for light-duty vehicles
   Investment in rail versus transport
   Energy intensity of freight transport

Note: Indicators in italics are performance indicators. The rest are policy indicators.
Source: ACEEE, *International Energy Efficiency Scorecard*, 2018.

Table 4.2: Top Five Countries in the ACEEE *Scorecard*, by Category

National Effort	Buildings	Industry	Transport
Germany	Spain	Japan	France
United Kingdom	France	Germany	India
Italy	United Kingdom	Italy	Italy
Japan	Netherlands	United Kingdom	China
France	Germany	France	United Kingdom

Source: ACEEE, *International Energy Efficiency Scorecard*, 2018.

The bottom five in the *Scorecard* – the countries leaving the most 20-dollar bills on the ground – were Thailand, Brazil, the United Arab Emirates, South Africa, and Saudi Arabia. The world's largest economy, the US, was eighth in national effort, doing well due to having extensive national appliance standards; it did less well in buildings, industry, and transport. A bit of good news is that energy intensity improved in all countries but Brazil from 2010 to 2015 (Figure 4.1).

Examples of best practices in each category highlight what leading countries are doing. EU members do well; they must follow the EU's Energy Efficiency Directive Targets, which call for (from 2008 levels) a 20% improvement in efficiency and a 50% reduction in energy use by 2030.[49] Under its 2014 National Action Plan on Energy Efficiency, Germany is mobilizing investment to renovate buildings and upgrade transport. Mexico's National Program for the Sustainable Use of Energy offers consumer incentives for replacing old products, along with subsidized homeowner and business loans. Best building practices are from France, with its innovative retrofit programs, and the US, which has strong building codes in many states and cities (but not in others), and, as noted above, does well in appliances and equipment. The US record on appliances and equipment, however, was stalled by the Trump administration's efforts to scale back these efficiency standards, a setback discussed in Chapter 8.

Japan leads in industry best practices. It has low energy intensity and a national policy for industry (The Act Concerning the Rational Use of Energy). In transport, the best-practice countries are France, which supports vehicle efficiency and mobility options, and Italy, with efficient passenger fleets and a higher investment in rail relative to road transport. (Rail is a more energy-efficient way to move freight.)

The *Scorecard* notes that, without the efficiency gains since 2000, the world would have used 12% more energy and endured substantially more emissions. Globally, in the leading countries, households saved 10% to 30% on their energy bills. Yet "energy efficiency remains massively underutilized."[50] Even the leaders "have substantial opportunities for improvement."[51]

Policies for Energy Efficiency

Government policy is central to the clean energy transition. Many existing costs are social and are not accounted for in markets and decision-making. Someone has to coordinate the many moving parts: investments, grid modernization, renewables expansion, physical infrastructure, and vehicle electrification, among others. Government policy is essential in speeding up the transition.

At first glance, efficiency should be the easy part. Using less energy saves money – a direct bottom line to benefit households, firms, and society. Payback periods often are short; efficiency compares well with other ways of cutting emissions. Yet even rational, profit-maximizing firms miss opportunities. The question then becomes how to use policy to promote enhanced efficiency.

Policies for accelerating the transition include mandates, financial incentives and support, market creation, and information disclosure. All of these policy tools may promote efficiency. This section presents a first brief look at these tools and how governments may organize them into strategies, while Chapter 8 will develop the topic in more detail.

Mandates that may be used to promote efficiency include building codes and appliance standards (including minimum efficiency performance standards, or MEPS), as well as ones imposed at a utility scale as energy efficiency resource standards (EERS). Given that a lack of up-front capital is a barrier to investing in efficiency, *financial incentives and support* have a role. Because information often is lacking, information *disclosure* has its uses. *Market creation* contributes directly (government procurement) and indirectly (carbon trading), but it is less prominent in efficiency policies.

Of course, some policies accelerate multiple elements of the clean energy transition, from efficiency, to renewables, to carbon removal. Cross-cutting tools attach a price to carbon through a trading system or carbon tax. Any policy that incorporates social costs into economic calculations promotes progress on multiple fronts. Like other decarbonization pillars, efficiency benefits from tools that are tailored to specific energy uses in transport, buildings, industry, and electricity. As with other clean energy goals, multiple policies should be combined to get the best results.[52]

Business is not averse to well-designed regulation. In the Economist Intelligence Unit survey cited earlier, 75% of executives said efficiency regulation "benefits the building sector" by "leveling the playing field, thus strengthening the case for investments."[53] This reinforces longstanding arguments in the field of environmental regulation that well-designed regulation supports business efficiency.[54]

Energy Efficiency and Decarbonization

More than other decarbonization pillars, energy efficiency is a no-brainer. True, wind and solar prices are falling, and the economic case for them is compelling. But efficiency investments typically pay off quickly and show up in the bottom line. The public interest in using energy efficiently is clear: Every dollar saved from the bottom line pays off in reduced pollution, less ecological damage, and less harm to health. In practical terms, efficiency allows us to decarbonize more rapidly than building new renewables plants to generate electricity or installing carbon capture and storage technology. It truly is the *first fuel*.[55]

Accepting the inevitability of economic growth – and the moral imperative to ensure a better life in poor countries as well as more equity in affluent ones – does not mean we should not think about the relationship of growth to quality of life. Still, the task here is to take the world as it is and consider how to decarbonize. Using less energy, or not using more later in this century, is part of the transition to a clean energy system. The world will use energy, and it will have to come from sources other than coal, oil, and natural gas. Renewable energy is the second pillar of decarbonization and subject of Chapter 5.

Guide to Further Reading

American Council for an Energy Efficient Economy, *The International Energy Efficiency Scorecard*, 2020.

Marilyn Brown and Yu Wang, *Green Savings: How Policies and Markets Drive Energy Efficiency*, Praeger, 2016.

David Goldstein, "Renewables May Be Plunging in Price, But Efficiency Remains the Cornerstone in a Clean Energy Economy," *The Electricity Journal* 31 (2018), 16–19.

Xavier Labandeira et al., "The Impacts of Energy Efficiency Policies: Meta-Analysis," *Energy Policy* 147 (2020), 1190–208.

5

Endless Flows

Renewable Energy

Anyone looking for evidence of a clean energy transition would find hope in a headline from *The Guardian* in June 2020: "Renewables Surpass Coal in US Energy Generation for the First Time in 130 Years." Renewables – wind, solar, hydro, geothermal, and biomass – exceeded coal for the first time since 1885, "when Mark Twain published *The Adventures of Huckleberry Finn* and America's first skyscraper was erected in Chicago."[1] This headline referred to 2019, but coal had been declining for years. This was the sixth consecutive year of its decline, resulting from an increased use of natural gas and renewables. The EIA projected that coal, which generated half of US electricity as recently as 2009, would fall below 20%, the lowest in four decades. As one analyst put it: "We are seeing the end of coal."[2]

Yet coal is not going away, at least outside of countries with mature economies. It may be declining in Europe and the US, but it is a mainstay in Asia's rapid-growth countries. They need prodigious amounts of energy to fuel industry and support a rising middle class. The rich countries have used up their hypothetical carbon budgets, and now emerging economies want to draw on their own accounts. Even in developed countries, coal still is used for electricity and in industry.

Renewable energy includes both the oldest and the newest forms of energy. They are old because they have always existed and were the first energy sources put to human use: sunlight for warmth, wind for mills and sails, water for grinding wheat into flour. For many reasons – energy density, convenience, portability, and cost – fossil fuels replaced them. Now we are going back to the future and replacing fossil fuels with renewables, only this time with help from modern technologies.

The rise of wind and solar has been one of the pleasant surprises of the energy transition. In 1998, the IEA's *Global Energy Outlook* did not even bother to list them. Renewable energy meant hydropower. The IEA predicted that non-hydro renewables "will still represent less than one percent of world electricity generation by 2020."[3] Similarly, in its 2000 *International Energy Outlook*, the EIA noted "it is difficult to foresee significant widespread increases in renewable energy use," adding, "It remains unlikely that renewable energy can compete economically over the projection period [to 2020]."[4] Fortunately, those projections were well off the mark.

The clean energy transition of the twenty-first century will occur on the back of renewables, mainly wind and PV solar. Prices are falling rapidly, especially for PV. According to the IRENA, between 2010 and 2020, the LCOE for utility-scale solar PV fell by 85%, onshore wind by 56%, offshore wind by 48%, and concentrated solar power (CSP) by 68%. Bioenergy/biomass was stable, while hydro and geothermal increased due to the costs of accessing more difficult sites, although hydro especially is still competitive (see Table 5.1). PV and wind costs will continue to fall. Meanwhile, the US National Renewable Energy Laboratory (NREL) projects that the capital costs for lithium-ion four-hour batteries may fall from 2019 levels by 26–63% (in 2030) and 44–78% (in 2050), to as low as $88 per kWh.[5] Calculated by LCOE, four-hour batteries in 2019 were at about $156 per kWh, 86% below 2010, and are likely to fall another 60% by 2030.[6]

Economies of scale and experience play a big role in these cost declines. Bloomberg New Energy Finance (BNEF) estimates that for every doubling in capacity the price of PV modules will fall by 28%, wind turbines by 14%, and lithium-ion batteries by 18%.[7] In two-thirds of the world, wind and solar are the least expensive option for new electricity generation; by 2030, they will be cheaper than oil and gas "almost everywhere." Europe will be the fastest region to decarbonize, with renewables accounting for 92% of its electricity by 2050.

Figure 5.1 shows the cost declines in generating electricity over the past decade for onshore/offshore wind and PV/concentrated solar (all described later). These declines will continue. In 2020, the IRENA stated: "renewables steadily increasing competitiveness,

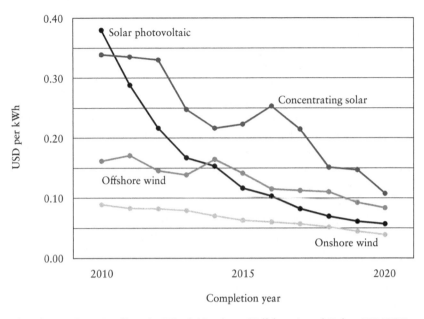

Figure 5.1: Cost Declines in Wind (Onshore/Offshore) and Solar (PV/CSP)
Source: Wikimedia Commons, from IRENA, *Renewable Power Generation Costs in 2019* (2020), https://commons.wikimedia.org/wiki/File:2010-_Cost_of_renewable_energy_-_IRENA.svg.

along with their modularity, scalability, and job creation potential, make them highly attractive."[8] Wind and solar may generate 60–80% of electricity later this century – promising news for clean energy.

Again, however, clean energy is a good news/bad news story. Although the amount of energy from renewables grew far more than expected, it has not gone up much as a share of total energy. Recall that fossil fuels still account for over 80% of total primary energy use, roughly the same as in the 1980s. The reason is that the world consumes more energy each year; renewables make it less dirty than it would have been but not necessarily cleaner overall. This fact underscores two critical conditions for a transition. The first is the need for steady efficiency gains. Despite falling energy intensity, demand goes up each year (the 2020 pandemic excepted). Second, wind and PV must continue to grow rapidly and at the needed scale, as must capacities for chemical battery storage.

This chapter looks at renewable energy's role in the quest for decarbonization. We begin, as in other chapters, with some basics. Although wind and solar carry most of the load, other renewables play complementary, sometimes niche, roles. Although their costs are falling, a fact that makes the energy transition possible, certain features of renewables make decarbonizing a challenge. These have to be accounted for. After that, the chapter considers legal and policy issues and opportunities for innovation, followed by two brief cases of country-level experiences with renewable energy.

The Basics of Renewable Energy

Renewable energy "is not substantially depleted by continued use."[9] The world will not run out of wind, sun, waves, or tides. It will someday run out of fossil fuels; coal reserves may last another century or so. Oil and natural gas, which are increasingly difficult, risky, and expensive to get out of the ground or from deep reservoirs offshore, could be depleted by later in this century.

Of course, it is not worry about depleting fossil fuels that drives the clean energy transition, it is the environmental and health impacts. Sources like wind, solar, geothermal, hydro, wave, and tidal do not emit CO_2 and other pollutants, at least in operation. They do not risk catastrophic oil spills, require blowing off mountaintops, cause leaking pipelines, or emit health-damaging particulate matter. Renewables have social costs, but they are far fewer than those from fossil fuels. Public health alone makes a case for renewables. Falling costs and far less climate impact are icing on the cake.

Not being exhaustible and having fewer social costs are not the only reasons renewables are growing rapidly. There are no perfect options when it comes to energy, but renewables are less imperfect than the alternatives. Onshore wind and PV now have *grid parity* with coal and gas in most of the world, and compete well in market price.[10] Geothermal is limited geographically but is competitive in many parts of the world. Offshore wind has many advantages, and costs will fall. Concentrated solar power is becoming competitive and has strengths in energy storage.

In addition to low social costs and competitive prices, renewables offer price stability. Most wind and solar costs consist of the capital and related expenses involved in building plants. Once they are operating, there are no fuel costs. The sun and wind are not metered. In contrast, fossil fuel prices rise and fall with markets. Renewables also offer flexible scales of operation. Solar PV is done at a utility scale, where lots of modules generate power for thousands of households and businesses; or it serves small numbers of users through residential solar and microgrids, all forms of distributed generation. Wind is not as flexible as PV but is less costly. It can operate at utility and smaller scales. Distributed generation is an advantage, especially for electricity access in developing countries.

One distinction among renewables will be central to my later discussion of their integration into electricity grids. Wind and solar are sources of *variable* generation and not always available. In contrast, hydropower, geothermal, and biomass are *firm* renewables, akin to dispatchable fossil fuels.[11] These distinctions are not always clear cut, but they matter in managing reliable grids. This section describes each renewable. A summary of costs and capacity factors is in Table 5.1 at the end of this section.

Solar Energy

Energy from the sun is useful in two forms. Solar photovoltaic cells convert solar energy directly to electricity: "If one were asked to design the ideal energy conversion system, it would be difficult to devise something better than the solar photovoltaic (PV) cell."[12] Although the scientific basis for PV technology goes back to the mid-eighteenth century, the practical applications for converting sunlight into energy date from work done at Bell Labs in Edison, New Jersey, in the 1950s. PV takes the original source of all energy – the sun – and turns it into useful energy as electricity.

The sun also delivers energy via concentrated solar power. While PV technology converts sunlight directly through a chemical process, CSP uses mirrors and lenses to concentrate solar energy, generate steam, and drive turbines, much like conventional coal and gas plants. This is only one application of a broader category of solar thermal energy, which includes such uses as passive space heating in buildings and solar collectors to heat water. Because the emphasis here is on large-scale electricity production, my focus is on concentrated solar and especially PV. Yet solar thermal energy more generally will play a role in the transition, especially in space and water heating. Of the two solar applications, PV has seen the most rapid growth and cost declines. Costs for both will fall in the coming years. CSP has an advantage that almost guarantees it a role: used with molten salt technology, it is more feasible and less costly for energy storage.[13]

Still, PV overwhelmingly is the solar option of choice at scale. As of the end of 2018, the installed capacity of PV was eighty times that of CSP. This is due to its phenomenal growth, which between 2009 and 2018 was more than twenty-fold. The LCOE of solar PV has fallen more than any other energy technology. In 2018 alone, PV installed capacity grew by 24%. Still, it was less than 1% of global capacity and 3% of renewables.[14] Bear in mind that this growth started at nearly zero. The PV leader is China, followed by the US and Germany. Other countries rapidly growing in PV use are India, Italy and the UK. Per capita, the hands-down leader is Germany, with about 1% of world population but nearly 20% of PV capacity. An issue with PV is its low capacity factor, which improved marginally from 14% to 16% in the last decade (Table 5.1)

Wind Energy

Wind energy also comes in two forms; the difference lies not in technology but in setting.[15] Turbines convert the kinetic energy of wind into electrical energy. Until recently, the primary uses of wind turbines have been on land, for onshore production. Onshore costs fell steadily over the last two decades (although not as rapidly as PV), at some 48% between 2010 and 2020 (see Table 5.1). This accounts for the dramatic increases in wind generation in such areas as the American plains, where states like Iowa, Kansas, and the Dakotas generate a large share of electricity with wind turbines. The largest US wind-generating site is in Wyoming, the top coal-producing state.[16] Onshore wind had a capacity factor of about 40% in 2020, slightly better than in 2010 (Table 5.1).

Offshore wind generation is more recent, and the costs are higher. This reflects the complex engineering required for offshore sites, the materials needed in corrosive environments, and the higher connection costs. Offshore wind does, however, have its advantages. Higher windspeeds give offshore sites better capacity factors; a good location on land can reach 30–40%, but a good offshore site may reach 40–50% or more. This affects the cost of the technology and will help close the price gap.

As of the end of 2018, onshore sites accounted for some 96% of global wind capacity. The leaders here are China, the US, Germany, India, Spain, and the UK.[17] Wind capacity almost quadrupled from 2009 to 2018; this is rapid growth but not on the PV level. In recent years, wind has been expanding about 10% annually. Wind and PV accounted for 84% of new global electricity capacity in 2018, a sign that they are the workhorses of the transition. Per capita, Denmark leads in wind. Its success is explored later. Offshore wind is being pursued in Europe, especially in the UK and Ireland, which have some of the best wind resources in the world. Offshore wind is drawing interest in the US, stimulated by the Biden administration's energy policies.[18] With advances in designs and technology for floating wind platforms, higher windspeeds offshore are becoming more accessible.

Hydropower

Until recently, renewable energy was largely limited to hydropower. Hydro uses kinetic energy from flowing water to drive turbines that generate electricity. Generating facilities come in many sizes, from small micro-generators to huge ones like China's Three Gorges Dam, with its thirty-four generators and 22,500 MW of capacity. In 2018, hydro accounted for some 40% of renewables globally, but with lower growth rates than wind and PV.[19] Costs have increased with more difficult siting.

China is the top hydro-generating country in the world; next are the US, Brazil, and Canada. Extensive hydro capacities in Brazil and Canada make them perennial world leaders in renewables. Per capita, Norway is the top hydro generator. This makes Denmark's reliance on wind possible, because it imports hydro-generated energy to offset variable wind supplies. Brazil, Canada, and Norway, with 3% of the world's population, account for 17% of global hydro.

Although marginal efficiency gains in hydro may occur, there is limited potential for improving core technologies, and it is seen as a mature technology. The major hydro issues are its environmental impacts and, relatedly, a lack of sites for expansion. Even in the aggressive WWS scenario discussed in Chapter 3, hydro accounts for only 4% of electricity by 2050. This is a likely high end. Yet it has a role. High capacity factors make it useful for baseload generation, and it is low-cost at good sites. It can also be ramped up quickly and cost-effectively, so is good for backing up wind and PV variability.

A variation on hydro power is a mainstay of energy storage. The most common form of storage in the world just now (about 94% of the total) is *pumped hydro*. Like all storage, it is a form of time shifting.[20] When excess variable wind and PV supplies exist, water is pumped to a high elevation and released when needed, using gravity to drive turbines to generate electricity.

Geothermal Energy

Geothermal technologies take heat from the earth and use it directly as a heat source or indirectly to drive turbines that generate electricity. Our focus is the second use, although direct uses for heating also contribute to clean energy. Unlike other renewables, geothermal is not based on the sun's energy. Nor is it fully renewable, given that overused sites may be depleted over time. Still, it is renewable insofar as it provides a steady flow of available, clean energy. It enjoys the highest capacity factor among renewables, at over 80% in 2020, on a level with fossil fuels and nuclear. Geothermal costs have gone up largely due to the expenses of accessing difficult sites.

Geothermal takes advantage of temperature differences between the surface and interior of the earth. Hotter interior temperatures are the product of "tiny quantities of long-lived radioactive isotopes … which liberate heat as they decay."[21] Geothermal technology uses drilling techniques similar to those in oil and gas production. There are no fuel costs, and operating costs are modest. Most costs are upfront capital and related expenses. Although costs vary by location, good sites can generate power cost-effectively.

The advantage of geothermal is that it is a *firm* source – there when we need it. Well-sited and well-designed plants have high capacity factors and are reliable for baseload power. They do require certain geological conditions. The best development areas are in the Pacific Rim – western North and South America, eastern Asia – and select areas like Iceland. Top geothermal countries are the US, the Philippines, Indonesia, Mexico, New Zealand, Iceland, Italy, Japan, El Salvador, and Kenya. Still, geothermal generates well under 1% of world electricity.

Ocean Energy

Any repetitive movement on a large scale is a potential energy source. Tides and waves fit that bill and are on the list of renewables. Since energy from the sun is called solar, energy from tides could be termed lunar. Tides result from the gravitational pull of the moon (and, less so, of the sun) on oceans. Wave technologies extract energy from the action of the wind on water. They work similarly to hydro in generating electricity from moving water, although differing in technology and scale.

Ocean sources (tidal and wave) generate a fraction of 1% of global electricity, are used selectively, and are expensive. They are a proverbial drop in the global energy bucket. They are listed here because they have a modest future role, offer endless energy flows, and are clean. Tidal technologies place a *barrage* (a low dam) across an estuary and release water to drive turbines or extract energy directly from tidal currents. Wave technologies capture kinetic energy from moving water to drive turbines. Tides deliver twice-daily energy bursts and are variable but predictable. Waves offer a steadier source and may back up wind and solar. Some experts think ocean sources may someday generate up to 15–20% of electricity in places with ample resources, such as the UK.[22]

Whether or not ocean energy can be scaled up later this century is an open question. Costs are high, but so were wind and solar decades ago. And the technologies can be tailored to meet particular needs. For ocean energy, this could include the job of providing electricity on remote islands or coastal areas that now rely on diesel fuel. And it could meet needs for special purposes: desalination, water pumping, and producing green hydrogen.

Energy from Biomass

Excepting geothermal and tidal, renewables derive energy from the sun. Biomass, referring to a diverse bunch of sources, also starts with the sun. It includes "organic materials such as wood, straw, oilseeds or animal wastes which are, or were until recently, living matter."[23] There is a difference between the traditional biomass used in poor countries and modern biomass, which is my focus here. This section considers biomass used to generate electricity. Liquid biofuels for transport are discussed in Chapter 6. Biomass exhibits capacity factors in the range of 60% to nearly 90%, depending on the form and combustion technology.

Until the 1800s, biomass was the world's main energy source. An early transition replaced wood and other materials with coal during the industrial revolution. The status of biomass as a renewable is questioned, because it includes materials and processes that are not all that green. Experts distinguish *beneficial* and *harmful* biomass to mark those with negative environmental impacts.[24]

Plant biomass comes in many forms: woody (trees), cellulosic (grass or seaweed), starchy and sugary (cereals, sugar cane), oily (palm oil), and microalgae. Wood is the largest source for electricity and for building and industry heat. Other biomass includes agricultural crops and waste, animal manures (e.g., chicken litter), human waste, and general municipal waste. In 2017, biomass accounted for nearly 10% of global electricity from renewables, behind hydro and wind. Yet it dominates heat production and comprises 96% of renewables in industry, building heating, and other uses.[25]

An illustration of the controversy over biomass is using forest wood and woody residue to generate electricity. Like other plant-based biomass, this is considered to be renewable "because its inherent energy comes from the sun and because it can regrow in a relatively short period of time."[26] In many US states, wood qualifies as renewable under clean energy standards, and federal tax preferences apply. Yet critics question this status. Wood may not be replenished as fast as it is removed, and it emits CO_2. Its renewability varies by location, forest practices, and combustion technology. Wood products are not always carbon neutral or fully renewable. Experts argue for distinguishing "between the types of biomass energy that are beneficial and those that are detrimental."[27]

Table 5.1 gives a summary of costs and capacity factors for renewables in the last decade. It describes the impressive cost declines in solar PV, CSP, and both forms of wind. Costs for bioenergy (biomass) vary by source material and overall have remained stable. Because of the challenges of more difficult siting and using less productive sites, hydro and geothermal costs have increased.

Table 5.1: LCOE and Capacity Factors for Renewables, 2010–2020

Technology	2010 LCOE	2020 LCOE	Change 2010–2020	2010 Capacity Factor	2020 Capacity Factor	Change 2010–2020
Solar PV	.381	.057	-85%	14%	16%	14%
Onshore wind	.089	.039	-56%	27%	36%	33%
Offshore wind	.162	.084	-48%	38%	40%	5%
CSP	.340	.108	-68%	30%	42%	40%
Hydro	.038	.044	18%	44%	46%	4%
Bioenergy	.076	.076	0	72%	70%	-3%
Geothermal	.049	.071	45%	87%	83%	-5%

Note: The levelized cost of electricity (LCOE) is given in US dollars per kilowatt-hour. Costs are the global weighted average for newly commissioned projects. Ocean sources are not included. Percentage changes are rounded to the nearest whole number.
Source: IRENA, *Renewable Power Generation Costs in 2020*, https://www.irena.org/publications/2021/Jun/Renewable-Power-Costs-in-2020.

The Environmental Impacts

A decided advantage of renewables is that they are *clean*. Most of the time, when we talk of them as clean energy, we mean they emit no CO2 or air pollutants like particulate matter and nitrogen dioxide. As discussed in Chapter 2, they are also ecologically friendly, relatively benign for ecosystems, wildlife, and water resources, and unlikely to cause catastrophic events such as oil spills.

From an environmental point of view, however, there is no flawless energy source. All are problematic. Wind and solar use land and have visual impacts. Hydro disrupts ecosystems and harms aquatic life, and large projects displace communities. Ocean sources may affect local ecosystems. Some biomass has impacts that probably should disqualify it as clean energy. The idea that there is a *perfect* energy source is an illusion. The standard should be the effect of meeting needs one way rather than another. The place to start is efficiency. The last resort should be carbon removal. In between, renewables are far preferable to fossil fuels, but they are not without negatives.

Critics claim that the problem with wind and solar lies not in their operation but in upstream costs – that materials and construction impacts undermine their greenness. This criticism took the spotlight in debates over the accuracy of Michael Moore's 2020 film *Planet of the Humans*. One claim in the film is that, although renewables have fewer emissions once they are operating, those from materials and construction offset that advantage. This is one of many claims in the film that drew criticism. Although there may have been some validity to this argument a decade earlier, there was very little by 2020. As far back as 2013, the National Renewable Energy Laboratory, part of the US Department of Energy, reviewed 2,100 studies of life-cycle greenhouse gas emissions from wind, PV, CSP, biomass, geothermal, ocean, hydro, nuclear, natural gas, and coal. It found that life-cycle emissions from renewables and nuclear "are much lower and less variable than those from fossil fuels."[28] In contrast, over its life cycle, coal "releases about 20 times more GHGs per kilowatt-hour than solar, wind, and nuclear technology."

Renewables cause far less health and ecological damage than fossil fuels. Nonetheless, we should be aware of their negative effects and minimize them. Most harm is in manufacturing, not operation. PV production, for example, uses toxics that can affect workers; those materials eventually must be recycled or disposed of. Wind turbines can harm birds or bats and cause low-level noise. Hydro and biomass raise a range of social and environmental issues. Big hydro projects displace communities, alter ecosystems, and harm aquatic life. Growing biomass pollutes water and harms wildlife. Table 5.2 lists the principal social costs of renewable sources and the actions that could minimize them.[29]

Wind and solar require land. One study found that coal, gas, and nuclear needed just under thirteen acres per megawatt of electricity generated; solar used over three times and wind nearly six times as much land in its full life cycle. Hydro needed more, over five times that of wind.[30] Another study estimated that meeting 100% of US electricity demand in 2050 with biomass would require 33,000 square kilometers of land (more than Massachusetts), or 12,000 if concentrated in sunny areas. (For comparison, the US uses 10,000 square kilometers on golf courses.) Among renewables, PV uses the least land through its life cycle and biomass the most. Still, once sites are operating, there is little land disruption and no resource extraction for wind and solar. Plants may be sited on low-value, multi-purpose land: farmland, industrial sites, brownfields.[31] Acreage may be less of a barrier than siting. Samantha Gross notes that the

Table 5.2: Environmental Impacts of Renewable Energy Sources

Renewable Source	Environmental & Social Impacts	To Minimize Impacts
Wind	Land use & ecosystem impacts; disrupting habitat; wildlife impacts (birds); noise; visual impacts.	Combine with low-value use (brownfields); recycle materials; use good locations.
Solar	Water use in manufacturing & (for CSP) operation; hazardous materials in construction; land use and habitat impacts.	Combine with low-value use; recycle materials or alternative designs.
Hydro	Ecological impacts; harm to aquatic life; possible methane releases; disruption of communities.	Locate & design sites to minimize impacts; limit applications.
Biomass	Depending on form of biomass, possibly: land use; water quality impacts; water uses; fertilizer harm.	Use beneficial biomass and avoid harmful forms; smart farming practices.
Geothermal	Land disruption for drilling; possibly air pollution; some potentially toxic sludge.	Use good sites and closed-loop systems.
Ocean	Impacts on marine life; ecological damage; interference with fishing & shipping.	Smart siting and good designs.

Source: Union of Concerned Scientists, *Environmental Impacts of Renewable Energy Technologies*, https://www.ucsusa.org/resources/environmental-impacts-renewable-energy-technologies.

"inherent attributes of wind and solar generation make conflicts over land use and project siting more likely."[32]

The life-cycle impacts of renewables have to be accounted for. Still, we can say that most emissions and other environmental impacts of renewables are in materials and construction rather than in operation. A life-cycle analysis supports a preference for renewables, depending on such issues as their siting and design, materials used, and the recycling rates of materials critical to operations.

Constraints on Wind and Solar Deployment

Falling costs remove one constraint on the use of wind and solar. Grid price parity is a ticket to competing in electricity markets. Coal's decline means it often is less competitive, a boost for wind and solar. Natural gas is competitive, although costs could grow over time. Experts see a role for natural gas in some applications until late this century. The issue is how much.

Yet costs are not the only issue. As we have seen, wind and solar variability is a challenge. Coal, gas, and nuclear are there when needed, with high capacity factors. Wind and solar do not share that same availability. Combine the electricity demand variability with the supply variability, and the challenge is clear. Another issue is moving wind and solar electricity from generation to point of use. The best wind and sun resources and land for siting may not be located where it is needed. This is a plus for offshore wind, with most population centers located along coastlines. Wind and solar also face other challenges: the need for critical materials and land, and difficulties in siting. These issues are introduced here and developed in more detail in Chapter 6.

Coping with variability

A decarbonized world will be far more electrified, hence integrating wind and solar into electricity systems reliably and cost-effectively is essential. As wind and solar grow, so do the challenges. Studies suggest that working 20% to 40% wind and solar into grids is doable; getting beyond 50% is harder. Moving past 80% or 90% defines the *really hard stuff* in the transition (see Chapter 6).

Variability poses two specific challenges to a broad scale-up of variable renewables. Both affect costs. Suppose a plant delivers electricity one-fifth of the time (the capacity factor for a good PV plant). In that case, it costs more per unit of energy, other things being equal, than one delivering four-fifths of the time or more – a typical capacity factor for coal, gas, or nuclear. That wind and PV have grid parity in most of the world shows they can overcome this challenge. Still, it means that as we move to a higher percentage of wind and solar, we need extra generating capacity just to meet demand reliably.

According to the Electric Power Research Institute (EPRI), an electricity system based entirely on wind and PV without complementary technologies like nuclear and CCS being part of the supply portfolio would involve "significantly higher" costs.[33] Another study found that electricity systems built on wind and solar must be "physically larger, requiring much greater installed capacity."[34] Installed capacity must be over four times the peak load (the highest level of demand) to be reliable. These are the issues keeping grid managers up at night. Because the cost of energy is such an important consideration in the transition, these issues have to be accounted for.

The next chapter, on electrification, examines these challenges in more depth. The point for now is that the wind and solar variability is a challenge that somehow has to be dealt with.

A second issue applies especially to PV. The genius of renewables, besides their being cleaner, is that once plants are operating, electricity is inexpensive. Reduced electricity demand during the 2020 pandemic illustrated this. Already declining in many parts of the world, coal plants suffered from falling demand because the marginal costs of each additional unit of PV were lower. In the UK, the cradle of the coal-enabled industrial revolution, solar reached an historical high on April 20, 2020, meeting 30% of electricity demand.[35] For a week in that same month, solar supplied 23% and wind 17% of Germany's electricity.[36] Once operating, PV and wind offer cheaper electricity. This is an advantage, but it leads to the second challenge: PV may be *too cheap* at times.

With excess capacity in the middle of the day, high levels of PV lead to **value deflation**, where "the economic value of additional wind and solar capacity decreases as their penetration increases."[37] Producers give away PV-generated electricity or sell it cheap. The more wind and PV generate at their peak, the less their marginal value. Storage helps, but suffers from similar issues. Varun Sivaram, a solar advocate, worries that value deflation "could plunge solar's value below the gently falling cost of producing it, thereby undermining the benefits of adding more solar power."[38] He adds: "This value deflation problem could halt solar's rise down the road. It's a counterintuitive penalty that the industry will incur for being too successful at adding solar capacity."[39]

The options at peak times, when PV and wind supply exceed demand, are to store it, sell it cheaply, give it away, or *curtail* it

– reduce the amount of electricity delivered to grids. A 2019 study, "Sunny with a Chance of Curtailment," found that grids may operate reliably with up to 55% annual PV generation, "but this would require high levels of curtailment and energy storage."[40] Details are discussed in the next chapter. What matters here is that levelized costs are not the only factor to consider for wind and solar. With value deflation, a question to ask is: "How can electricity markets ensure revenue sufficiency with generation that is essentially free?"[41] Giving things away is not a sound business model, and energy is a business, even if it is clean.

Moving energy to where it is used

Imagine the potential solar resources from the Sahara Desert in North Africa. Each square meter of land receives the equivalent of 2,000–3,000 kilowatt-hours of energy annually. With 9 million square meters, that comes to 22 billion gigawatt hours annually, according to the US National Aeronautics and Space Administration, enough to power Europe 7,000 times over.[42] Of course, this is a *technical*, not an *economic* or *practical* potential, but it describes a formidable source near densely populated Europe. A private firm, Desertec, tried to finance a Sahara project, but it collapsed in 2014.[43]

Many of the best solar resources are located in relatively unpopulated areas: the Sahara, the Middle East, the southwestern US, northern Chile, and Australia. A challenge for integrating solar into electricity systems is moving it from generation sites to population centers reliably and cost-effectively. The same holds for onshore and offshore wind. In the US, the prime onshore resources are in the Great Plains, where population densities are low. Texas has the most installed wind capacity of any US state, nearly three times that of second-place Iowa. If it were a country, Texas would rank fifth in the world in wind capacity. A reason is that Texas has "a boatload of wind," as one expert put it.[44] Another is that it invested in transmission capacity. Following a 2006 study, the Public Utility Commission of Texas created Competitive Renewable Energy Zones, aimed "at delivering electricity generated by wind farms in rural northern and western Texas to Dallas, Fort Worth, Austin, and other areas of the Lone Star State."[45] This is yet another illustration of the need to make all parts of the energy system work together – and of the need for a strong guiding hand.

Assessing National Progress on Renewables

My focus in this book is on country-level performance. This chapter has highlighted some leading performers, often the same countries that are doing well on efficiency. Germany ranks as a leader, with its policies for driving PV. Denmark is a top wind country. The UK is getting better at renewable energy, and has formidable offshore wind resources, especially off the coast of Scotland.

There are surprises: Kenya generates growing amounts of electricity from geothermal, where it has strong potential. Geothermal also gives Iceland one of the cleanest energy systems in the world.[46] Then there are the hydro stalwarts of Brazil, Canada, Norway, and Venezuela, who meet most electricity needs from flowing water. Spain moved rapidly on wind and solar, and South Korea deserves favorable mention for moving ocean energy forward, with good tidal potential.[47]

Quite a few systematic comparisons of countries on renewables and clean energy are available. The World Bank's *Regulatory Indicators for a Sustainable World* (RISE) uses law and policy indicators to assess 111 countries on three dimensions: efficiency, renewables, and electricity access. Indicators for renewables include overall legal and regulatory frameworks; state of planning for expansion; economic incentives and regulatory support; support for network grid connections; and carbon pricing policies. The 2017 version found "significant progress in developing enabling policy frameworks for renewable energy."[48] In 2017, 84% of countries had adopted a legal framework. A strength was having more renewable targets, stimulated by the Paris Agreement and EU directives. Yet many targets were not accompanied by sufficient supporting policies, especially for integrating renewables, promoting grid access, and overall renewable energy planning.

Many findings are consistent with what the IEA and others are saying. The electricity sector shows progress, "privileging it above the transport and heating and cooling sectors."[49] But a weakness is that only 26% of countries "integrate high-quality forecasting and grid flexibility assessment for variable renewable energy."[50] Transport, heating and cooling, and grid integration need attention. Germany earned first place in the RISE comparison, followed by the UK, Switzerland, India, and France. Rounding out the top ten were Italy, South Korea, Slovakia, Bulgaria, and the Netherlands.

Canada, Iceland, Sweden, Hungary, and Denmark also rated high. Rich nations led the list, although lower-income countries showed progress. Stand-outs among low-income countries were Ghana, Tunisia, and India. Most improved were Uganda, Malawi, Rwanda, and Jordan. Renewables in developing countries are critical because that is where most economic growth is yet to occur.

Other comparisons exist. One assesses countries in terms of their speed of scaling up. Excluding hydro, the fastest renewables growth from 2008 to 2018, according to the *BP Statistical Review of Energy 2020*, occurred in Egypt (up 83%), Japan (25%), Brazil, and Chile (over 17%). The US and China were at 13% and 15%, respectively. If we look only at the 2018–19 growth rate, China and Morocco led, up by more than one-third, while Chile expanded renewables by over 20%.[51]

Other comparisons are given in the *Climate Change Performance Index* (CCPI), which annually ranks fifty-seven countries and the EU, accounting for 90% of emissions. It rates climate and energy performance in four categories: greenhouse gas emissions (40% of the ranking), renewable generation (20%), energy use (20%), and climate policy (20%). Sweden was the top-ranked country in 2021, followed by the UK, Denmark, Morocco, Norway, Chile, India, Finland, Malta, and Latvia. Germany ranked nineteenth, the EU as a whole sixteenth, and China thirty-third. The US was dead last at fifty-seven, behind Iran and Saudi Arabia.[52] The US suffered in the policy ratings after four disastrous years under President Donald Trump. His administration reversed many of President Barack Obama's policies, weakened energy efficiency programs, expanded fossil fuel production, and withdrew from the Paris Agreement.

Given my focus on renewables in this chapter, that aspect of the CCPI is worth focusing on. The *Index* evaluated countries based on current levels and recent trends. Countries leading in percentage of renewables (including hydro) in total primary energy supply were Latvia, Norway, Sweden, Denmark, and Finland. Chile, a recent addition, was twelfth. Another indicator rated countries on their progress in scaling up non-hydro renewables. Countries rated *very high* included Turkey, Ireland, the UK, China, and Morocco.

Morocco is an interesting case. A country that historically imported nearly all of its energy, it recently set a goal of meeting 52% of its energy needs by 2030 with a mix of 20% wind, 20% solar, and 12% hydro.[53] Sweden is building on a record of leadership with diverse sources, and aims to eliminate fossil fuel electricity generation by 2040. The UK is scaling up wind, solar, and other renewables. China has the most installed wind and solar in the world, reflecting its status as the largest country by population and second-largest economy. Denmark, a small country with a fraction of a percent of China's population, has the highest proportion of wind-generated electricity in the world, at 47% in 2019 and with more each year.[54]

Using figures from Ember, a clean energy research group, Table 5.3 lists percentages of wind and solar electricity in selected countries for the first half of 2021. The world average is 10%, a number that may grow to 60–80%. Several countries are well above the world average: Denmark, Ireland, Germany, and the UK. Others are close to it, notably large economies like the US, China, Japan, and Brazil. The EU is leading the shift, at 21%. Russia has negligible renewables. Wind and solar generated almost as much electricity as nuclear, which stood at 10.5% for the first half of 2021.

As we saw in Chapter 2, renewables generate more jobs per unit of investment but displace workers in other sectors. China accounts for almost 40% of renewables jobs.[55] It leads in PV and wind, followed

Table 5.3: Percentage of Electricity Generated by Wind and Solar (first half of 2021)

Denmark	64%
Ireland	49%
Germany	42%
United Kingdom	33%
European Union	21%
Turkey	12%
Brazil	10%
China	10%
India	10%
Japan	10%
World	*10%*
Vietnam	6%
Canada	5%
South Korea	4%
Russia	0.2%

Note: Rounded up to the nearest whole number, except for Russia.
Source: Ember, *Wind and Solar Now Generate One-Tenth of Global Electricity* (August 2020), https://ember-climate.org/project/global-electricity-h12020.

by Japan and the US for PV and Germany and the US for wind. Brazil is a leader in liquid biofuels. Many countries are building export markets. Denmark accounted for 42% of wind exports, Germany 29%, and Spain 15%. China leads in PV, with nearly one-third of the total, followed by Japan (7%) and Germany (5%). The largest economy, the US, claims but 4% of PV exports and a mere 0.02% of wind. Work in the renewable energy industry calls for an array of skills and, the IRENA notes, "appeals to women in ways that the fossil fuel industry does not."[56] Women hold 32% of renewables jobs but only 22% of those in oil and gas. So there are gains for gender equity here as well.

Technology Opportunities and Growth Potential

Wind, geothermal, and biomass are relatively mature technologies. To be sure, we will find ways to improve their capacity factors and cost-effectiveness and to integrate them into regional and national electricity systems. For wind, the opportunities are in turbine designs, materials, efficiency, size, rotation speeds, and better wind forecasting. These offer prospects for incremental rather than radical technology gains. In contrast, offshore wind offers more upside technology and efficiency potential for reducing materials costs and in building and maintaining floating platforms.

Hydro is an even more mature technology. Under good conditions, it generates firm power cost-effectively. Yet large cost cuts are unlikely for hydro or geothermal. Hydro is constrained by limits in suitable sites and "by environmental concerns about large new dam projects."[57] Hydro depends on water resources affected by climate change, a future constraint. Biomass has so many forms that it is hard to generalize. The upside technology potential is limited, and some biomass uses should end. The likely advances for bioenergy lie more in transport, considered in Chapter 6.

Tidal and wave energy are nowhere close to large-scale commercialization. They play a small role in decarbonization scenarios, usually under 1% globally. They do, however, warrant investment for potential expansion and, especially, for meeting specialized needs. With innovation and economies of scale, their costs will fall. As two experts put it, "If wave energy costs follow the example of wind power then rapid cost reductions can be expected."[58] PV and wind show how the economics can change with the right technology and market conditions.

Among renewables, PV has the most upside potential. It is based on silicon wafers, and their declining price is what enabled rapid growth. In the future, advanced materials and wafer designs will offer higher conversion efficiencies. One potential innovation is adding layers to the wafers and using the full solar spectrum with materials like gallium arsenide. Another prospect is coating silicon with perovskites that enable wafers to draw on multiple parts of the solar spectrum.[59] Other prospects are drawing energy from indoor lighting, opening up all kinds of innovation potential.

Another constraint is reliable access to critical materials. Like any technology, renewables rely on specific materials available only in certain areas. Because these affect electric vehicles as well, they are considered in the next chapter.

Two Country Illustrations

Countries vary in their ability to work renewables into energy systems. Where there has been progress, it usually is in electricity. The least progress has been in transport, industry, and buildings, which will be discussed in the next chapter. Here we will focus on electricity in two country-level cases.

Denmark: gone with the wind

Many characteristics of global energy trace back to the effects of the 1970s oil embargos. Developed countries were forced to deal with their dependence on imported oil from often unstable or unfriendly parts of the world. The concept of *energy security* entered the lexicon. These external shocks provoked rapid changes: France moved into nuclear; Sweden and Japan doubled down on efficiency; Sweden later adopted a stringent carbon tax; Brazil expanded biofuel investments.[60] In contrast, clean energy progress evaporated in the US when prices and supplies recovered in the 1980s.

Denmark took its own path. A small country of under 6 million people, Denmark lacks fossil fuel resources. Until the 1970s, it relied on imported oil to run its economy, but the oil shocks forced a reckoning. Nuclear was not an option due to strong public opposition. What Denmark did have was a history of using wind resources and a topography and coastal features that are favorable to wind energy.

The transition was facilitated by the government's strategy. A major source of demand for wind energy came from agriculture. Policymakers and investors used this demand to gain support from farm communities by involving them in decision-making and ensuring that local agricultural cooperatives received a share of the profit. This built up local support and helped overcome opposition to siting facilities. This strategy, along with the other factors favoring wind, was successful. At the time of the second oil crisis in the late 1970s, Danish wind capacity was at 1 megawatt. By 1990, it had grown to 326 MW; by 2020 it was almost 6 *gigawatts*, with major expansions planned offshore and a goal of eliminating fossil fuels from the electricity sector by 2030.[61] Indeed, "the Danish case offers a picture of the feasibility of the renewable energy revolution."[62]

Another factor was access to Norwegian hydro. Norway has the highest hydro capacity in the world. As noted earlier, hydro offers a firm clean energy source backed up by variable sources like wind. Norway uses this to its advantage; as one journalist puts it, "Norway wants to be Europe's battery."[63] High-capacity transmission lines link Norway and Denmark. This balances Denmark's grid, despite wind variability. Denmark exports energy to Norway at peak supply times, and imports hydro when wind is unavailable. Yet there are limits, as argued by Jason Deign in "Why Norway Can't Become Europe's Battery Pack."[64] With a cold climate and electricity needed for heating, Norway has a large energy appetite that will grow as it expands its EV fleet and electrifies industry, heating, and cooling.

The case of Denmark demonstrates that a clean energy commitment can deliver results. Two aspects of its approach are worth noting. First, the driver for change was not climate and health but the country's dependence on imports at the time of the 1970s oil shocks. Second, its political strategy was to engage local communities and offer them benefits. These are useful lessons for other countries undergoing a clean energy transition.

Japan: getting back on track

In their 2018 book *Renewables: The Politics of a Global Energy Transition*, Michael Aklin and Johannes Urpelainen assess how external shocks serve as catalysts for change. Something has to punctuate the equilibrium. Illustrating carbon lock-in, existing carbon-based technologies remain dominant until a major event disrupts the status quo and stimulates a search for alternatives. If ever there was a shock leading to such disruption, it was the tsunami that shut down two coastal nuclear plants at Fukushima-Daiichi in March 2011.[65]

Fukushima raised two fears. One was nuclear safety. In 2011 nuclear plants generated over a quarter of Japan's electricity. Several events over the years had raised concerns about nuclear safety, among them Three Mile Island and Chernobyl. Another fear was over energy security. Because nearly all of Japan's nuclear plants were shut down after the tsunami, one-fourth of its electricity capacity quickly went offline. Japan lacks fossil fuels of its own. This should have set the stage for a transition.

A decade later, *The Economist* magazine reports, Japan still is waiting. In the short run, Fukushima had the expected effects. The government pushed a law supporting renewables through parliament. Investors started raising capital. Still, renewables generated only 17% of the country's electricity in 2017, up from 10% in 2010, and nearly half of that was old hydro, hardly a sign of a new era. Nuclear plants mostly were "replaced not by wind turbines and solar panels but by power stations that burn coal and natural gas."[66] By the late 2010s, coal, gas, and oil generated over 80% of electricity, much of it from imported Middle East oil and Australian coal. Nuclear fell from one fourth to 2% of electricity, although the plan is to build it back to 20% by 2030. As noted earlier, wind and solar in the first half of 2021 were about 10% of Japan's electricity, the world average.

By 2019, fossil fuels made up 88% of the primary energy supply. Emissions peaked in 2013 and since have dropped back to 2009, pre-Fukushima levels. Still, the country has a long way to go; in 2021, the IEA noted, "the carbon intensity of Japan's energy supply remains one of the highest" among developed countries.[67] Japan is working to get back on track, a decade post-Fukushima, with renewables, nuclear, CCS, efficiency, and hydrogen (which is discussed in the next chapter). The government has set a net-zero carbon target for 2050, like most developed countries.

Among the reasons for the sluggish change after Fukushima are geography and geology. Japan is mountainous; although it has geothermal and onshore wind resources, siting and transmission are difficult. Its electricity industry is not well designed or regulated in a way that allows flexibility for variable sources. An island with an isolated grid, Japan lacks access to a Norway for backup power. It did adopt feed-in tariffs in 2012, on the heels of Fukushima; this stimulated growth in renewables but was costly and requires other investment. As one expert put it: "as generation via renewables increases, various costs will be incurred for such needs as building more transmission lines, securing power storage and providing maintenance for backup thermal power generation."[68]

Even with the Fukushima shock, a clean energy transition is a challenge for Japan, as it will be for most large countries with a long dependence on fossil fuels. Its geography does not help. On the other hand, its population density and limited options are promoting a focus on hydrogen as a clean energy linchpin, one that could pay dividends for the rest of the world (see Chapter 7).

Renewables and Decarbonization

Renewable sources have clear advantages that make them central to decarbonizing. With a few exceptions, they are not depleted by use. They enhance energy security. Once operating, many deliver energy at low, predictable prices. Most important, renewables are better for health and the environment and have relatively low social costs. They are increasingly affordable. No technology is perfect, but renewables are better for the climate, health, and ecosystems than the alternatives.

Given their advantages, why does the world depend so much on fossil fuels? One answer is that fossil fuels perform well. They are energy-dense, deliver reliable energy flows on demand, are readily accessible (although less so every year), and offer (this also is changing) good returns on energy invested. Oil was the mobility fuel of choice because it is dense and portable. Once it became feasible (in the 1950s) to move natural gas long-distance via pipelines, it became a staple. With high capacity factors and availability, coal and gas were valued for generating electricity.

All these advantages were the foundation for carbon lock-in. Once they became dominant, fossil fuels and technologies acquired a life of their own. Complex systems are designed around them – think electricity grids and internal combustion engines – and consumers get comfortable. Investments in research and infrastructure focus on carbon technologies. Laws and policy make them preferred economically and tolerable environmentally. Indeed, despite their accomplishments, laws like the US Clean Air Act had the effect of sustaining carbon lock-in, making it manageable. Once they become dominant, carbon-based fuel interests acquire political and economic power that they use to block a shift to clean energy.

A second answer is that wind and solar only recently began to make economic sense. Wind has been on a half-century journey from its early uses in California in the 1970s, supplied by Danish turbines, to its uses in northern Europe in the 1990s, to the competitive source it is now.[69] PV has enjoyed faster growth in just the last decade. Despite their rapid expansion, onshore wind and PV have a long way to go. They began at a low base, and increased energy consumption has offset much of the growth.

All renewables can contribute, even if they are not global workhorses like wind and solar. Hydropower delivers baseload electricity in many parts of the world, backs up variable renewables, and meets special needs like powering desalination or producing green hydrogen. Geothermal may never amount to more than a fraction of global electricity, but is a mainstay in countries like Iceland, El Salvador, and Kenya. Biomass contributes in Sweden, Finland, and elsewhere, and helps in heat production. Wave and tidal have potential and warrant investment to support future scale-up.

Still, if renewables are not available at the needed scale, the transition will stall. Major growth must occur in onshore/offshore wind, solar PV, and CSP, and especially battery storage. Wind and solar dominate electricity generation in decarbonization scenarios. In some, they are nearly all of it. To produce on that level, they must be supported by battery storage as well as by enhanced and modernized grids, grid flexibility, forecasting, and other system-level capacities.

The next chapter takes up the third decarbonization pillar: electrification. A theme of this version of the transition is that we should electrify everything, or as much as feasible, to take advantage of wind, solar, and other renewables that produce clean electricity. This chapter introduced some issues associated with electrification. We may not electrify everything, nor do we have to, given likely progress on green hydrogen and other ways to carry and store energy. Still, a world powered by renewables may need to triple the amount of energy delivered as electricity by mid-century.

Guide to Further Reading

Michael Aklin and Johannes Urpelainen, *Renewables: The Politics of a Global Energy Transition*, MIT Press, 2018.

Stephen Peake, ed., *Renewable Energy: Power for a Sustainable Future*, 4th ed., Oxford, 2018.

Varun Sivaram, *Taming the Sun: Innovations to Harness Solar Energy and Power the Planet*, MIT Press, 2018.

Brice Usher, *Renewable Energy: A Primer for the 21st Century*, Columbia University Press, 2019.

6
Electrify Everything

In mid-August 2020, California suffered rolling blackouts (short-term interruptions in electricity supply) as it coped with an extreme, climate-change-related heat wave that led to spikes in electricity demand.[1] Critics of clean energy were quick to blame the blackouts on California's reliance on wind and solar. In fact, the causes were a complex combination of poor forecasting of next-day demand, miscues among providers, short-term loss of natural gas supplies that should have been available, and lack of electricity from other states also coping with extreme heat.[2]

The episode has lessons for clean energy. One is that extreme heat linked to climate change threatens electricity systems and other critical infrastructure and services. Aside from the sudden upswing in demand, it was reported that natural gas sources failed because their cooling systems could not keep up with the heat. A second lesson is that critics are ready to jump on renewables as a cause of problems, especially when it involves reliability. Third is how difficult it is to balance electricity supply with demand and to maintain grids. Wind and solar did not cause the outages; they did make the job of running grids more complex. These events also illustrate the need to avoid power failures in a transition. Nothing kills clean energy faster than perceptions that it is unreliable.

Electrify Everything, But Rely on Variable Sources

Late in 2018, three energy experts issued a review of forty decarbonization scenarios, like those discussed in Chapter 3. All of them expanded electricity use. Looking across the scenarios, the authors found that the electricity needed to meet demand by 2050 would have to increase 20–120%, and 120–440% by 2100. From making up 20% of global final energy in 2020, electricity would comprise 25–45% of delivered energy in 2050, and could be as high as 70% by 2100.[3] The reason for this massive increase, of course, is that electricity from renewables has to decarbonize transport, buildings, and industry. The transition starts with electricity and builds from there. Is this level of scale-up likely? How might it occur? What would a decarbonized electricity system look like? Could that system reliably electrify transport, buildings, and industry? Could it meet our needs and not cost too much?

A premise of the clean energy transition is that we have to *electrify everything*, or as much as possible. Wind, solar, and other renewables are close to ideal for generating electricity, but they are not suited to powering transport, *at least directly*. Indirectly, renewables generate electricity that powers vehicles. Or they make green hydrogen for cleaner transport. Or they produce synthetic fuels that are low in carbon. Until engineers come up with affordable wind- or solar-driven vehicles to move people and freight around at a huge scale, electricity is the key to decarbonizing transport. As we will see in this chapter, renewables can help directly, to a degree, in buildings and industry.

At first glance, the goal of electrifying everything might appear relatively easy to realize. After all, electricity use went from nothing in the late 1800s to a mainstay of the modern world in a matter of decades. Just as the transition from wood to coal enabled an industrial revolution in the 1800s, so electricity made prosperity possible, for much of the world, in the 1900s. Electricity is so vital to economic development that not having access to it is considered a major growth impediment.

The first part of this chapter considers electricity and its role in development. In the modern world, few commodities would be as sorely missed as electricity if it suddenly vanished. Lighting, appliances, heating and cooling, the internet, social media – it is hard to imagine a world without electricity. Indeed, what would life have been like during the Covid-19 pandemic without it? Now, with electricity being so critical to decarbonizing, it will be even more omnipresent and essential.

The chapter then turns to electricity's role in transport. Remember that transport accounts for over one-fourth of global emissions. It is the primary user of oil. Industry is another tough nut to crack. It accounts now for nearly half of emissions and will rise in rapid-growth areas of Asia and eventually Africa. Industry will not be fully electrified by later in this century. The last part of the chapter examines clean electricity's role in decarbonizing the *really tough* industry emissions.

Electricity and the Energy System

Electricity is an accumulation or movement of electrons to produce useful work – lighting, heating and cooling, driving machinery, powering appliances. Unlike coal, oil, or wind, electricity is not primary energy. It takes primary energy to produce electricity, which is a secondary source, an energy carrier. The ability to produce it goes back at least to Michael Faraday's experiments with electromagnetic forces in the 1830s. Electricity on a large scale was enabled by Thomas Edison's discovery of the parallel circuit in the 1880s.

This technology was applied at the first generating plant, the Pearl Street Station in Lower Manhattan, in 1883. Like other innovations of the era – the telephone and radio – it enables a separation of *space from time*, as Gretchen Bakke puts it in her history of the US grid.[4] Many experts consider electricity to be one of the major Kondratiev waves of technology change, this one occurring from the 1880s to 1930s, that made economic development possible.[5] Figure 6.1 shows the relationship of per capita electricity consumption to GDP per capita for 160 countries from 1980 to 2010. Electricity and GDP are tightly linked, with signs of *decoupling* due to increased energy productivity.

Electricity enables economic and social development; lack of access is a barrier to a better quality of life, especially in Sub-Saharan Africa, where electricity deficits are highest.[6] One study finds that, while "electricity access is likely not sufficient for economic growth, the data show that electricity use and GDP tend to go hand in hand."[7] While affluent countries seek to decarbonize, poor ones want to expand electricity access, ideally with clean sources.

Electrical grids are "one of the most impressive engineering feats of the modern era," and constitute "the largest machine in the world, a continent-spanning wonder of the modern age."[8] From its origins in

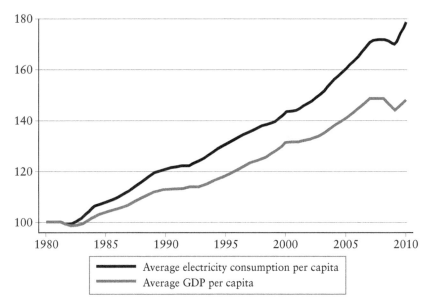

Figure 6.1: Relationship of Global Electricity Consumption and GDP Per Capita
Source: Fatih Karanti and Yuanjing Li, "Electricity Consumption and Economic Growth: Exploring Panel-Specific Differences," *Energy Policy* 82 (2015), 268. Data from 160 countries sourced from UNCTAD and EIA.

the late 1800s, the grid in much of the world evolved into a complex system of physical infrastructure, business relationships, technical standards, and policies to deliver power, most of the time, every second of every day, to billions of people around the world.

The grid in most places has four components. It starts with *generators* that produce electricity with coal, gas, nuclear, solar, wind, or hydro, and occasionally other sources.[9] Second are *transmission* lines that carry the electricity, often over long distances. Low-voltage power from generating plants is converted to a high voltage to reduce resistance losses in transmission. Third, transformers convert it back to a low voltage for *distribution* to homes, businesses, and other users. Fourth are *consumers*; their aggregate use makes up the electricity load, which varies daily and seasonally. The traditional grid in most countries relies on large centralized coal, nuclear, or natural gas plants that deliver power through elaborate transmission and distribution system.

Electricity is a commodity that is bought and sold in markets. It differs from other common commodities, such as corn or oil, in many respects. It cannot be stored easily or cheaply, and it has to be generated, transmitted, and distributed within strict technical limits. Most challenging of all, it has to be available on a flexible basis so that demand matches supply. Electricity demand (also termed total load) can be predicted *generally*, but not with any certainty. Among the sources of uncertainty is the weather, even in affluent areas, as was demonstrated by the Texas electricity crisis in 2021, caused by sub-zero temperatures.[10]

Electricity markets, their operations, and government regulation of them vary around the world, but they share features. The challenge for grid managers is to match supply with demand. Although they use long-term contracts to meet power needs, they also rely heavily on shorter-term, *day-ahead* auctions to purchase electricity. They usually meet over 90% of their needs with these auctions, but also do *real-time* purchases on an hourly or even more frequent basis.[11]

Load-serving entities like grids purchase the lowest-cost source available for meeting their needs. This is an advantage for many renewables, especially wind, because they have no fuel costs and can deliver electricity at low marginal prices. As demand goes up, grid managers call on more costly sources. Periods of peak demand involve the highest prices. It helps to minimize demand peaks – to even out daily and seasonal loads – in keeping prices low. And remember that these are markets – at least when electricity is deregulated, as it is in the EU and most of the US – meaning that areas with high demand and less supply from generation sources typically have higher prices.

Three aspects of clean energy represent a challenge for conventional grids: the expansion of renewables; the growth of distributed energy resources – residential solar, microgrids, and local storage; and deployment of information and communications technologies (ICT) like smart sensors and artificial intelligence. While these trends challenge grids, they also present opportunities. The main challenge is integrating wind and solar reliably and cost-effectively. Wind and solar are far cheaper than they used to be, as we saw in the previous chapter. This makes them feasible for doing more generation, but this is not accomplished easily. Having distributed resources means that electricity is available to more people – especially in developing countries where bulk grids are lacking – but it tests the old model. The opportunities come with the use of technology, particularly in the form of *smart grids*, but it takes money, planning, and policy to realize them. These challenges and opportunities will be our next topic.

The grid's job remains the same, however: deliver reliable, cost-effective electricity, on demand, to millions of households, businesses, schools, factories, and others.

Can Renewables Produce Electricity Reliably and at Scale?

As we have seen, rapidly falling wind and solar costs make the clean energy transition possible. These will continue to fall, as will storage costs, and both are critical in a renewables-based system.[12] Yet falling costs are only part of the picture. Like our sampling of decarbonization scenarios in Chapter 3, the 2018 study mentioned above identified two strategies. One relies heavily on variable renewables, sometimes almost entirely on wind and solar, along with substantial storage capacity. This is reflected in the 100% wind, water, and sun scenario. A second strategy is to draw upon a diverse portfolio of low-to-zero carbon sources, such as nuclear, biomass, and carbon capture and storage.

But why even think about problematic and controversial ways of generating electricity, like nuclear and some biomass, or the economically unproven (at a commercial scale) CCS? The issue lies with the *variability* (or intermittency) of wind and PV. Wind and sun are not always there when we need them. Their availability varies within 24-hour periods and seasonally. This is obvious with the sun. It rises and sets predictably and, in places far from the equator, is less available in colder months. Wind also varies, with typically more available at night and in the summer. Both vary in geographic availability, and this affects transmission networks as well as generation capacities.

Suppose wind- or sun-generated electricity could easily be stored for later use. In that case, this variability could be managed without too much trouble. Yet electricity cannot yet be stored easily, cheaply, or at the needed scale. As Bakke writes, "the electricity we use, day in and day out, is always fresh." In the case of wind, electricity was, just a minute ago, "a fast-moving-gust of air."[13] Without storage, electricity is yet another case of using or losing it. One side of the variability coin is periods when there is too little wind or sun to meet demand; the other side is when too much power is available and generation has to be *curtailed*.

On top of wind and solar variability are the fluctuations in daily and seasonal electricity demand. Figure 6.2 shows the shape of demand, the *daily load curve*, on a hot day in California in 1999. The top line shows the aggregate community demand; the other lines show the demand segments. It likely mirrors our daily routines. Demand is low late at night and in the early hours, bumps up during the working day, then *ramps up* in late afternoon and early evening. This would be even more challenging if commercial and residential demand did not offset each other slightly. The first ramps up during the working day, the second towards its end. This is when grids rely heavily on PV: as the sun sets and generation fades, demand ramps up, causing the *duck curve* as studied in California.[14] Demand also varies seasonally. Figure 6.3 presents the *annual* load curve for the

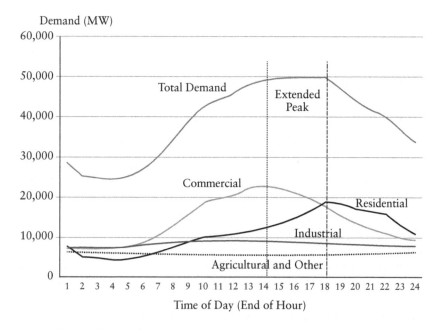

Figure 6.2: Daily Load Curve for a Hot Day in California in 1999
Source: *Electropedia*, "Battery and Energy Technologies." Based on data from Lawrence Berkeley National Laboratory, https://www.mpoweruk.com/electricity_demand.htm.

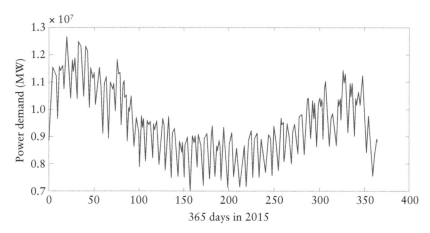

Figure 6.3: Annual Load Curve for the United Kingdom in 2015
Source: D. Li, W.-Y. Chiu, and H. Sun, "Demand Side Management in Microgrid Control Systems," in Magdi S. Mahmoud (ed.), *Microgrid: Advanced Control Methods and Renewable Energy System Integration* (Elsevier, 2017).

UK in 2015. In this case, demand is higher in cold than in warm months.

Electricity systems are finely tuned to match supply with demand. Having too little or too much power at any time literally throws them out of whack, leading to power failures or surges that damage equipment all along the line, from transformers to home appliances. Supply failures do not go down well with the public. Recurring failures will stall an energy transition, especially if they are perceived to be a result of using variable renewables. On the other hand, curtailing generation is costly. It undercuts wind and solar competitiveness with *value deflation*, discussed in Chapter 5, when marginal prices fall so low that they undermine the economic viability of a generating source.

How difficult is it to rely just on wind and solar? Here the perfect is the enemy of the good. The authors of the decarbonization survey mentioned earlier conclude that an electricity system based entirely on variable renewables (wind and solar) is "technically possible" but gets more difficult and costly as the proportion of variable sources increases and other options (say gas or nuclear) are eliminated. To rely entirely on wind and solar, a system would need "a continent-scale expansion of long-distance transmission capacity"; highly flexible demand management, to the extent of "reshaping demand to match variable supply" rather than the now-standard reverse practice; and "overbuilding total installed capacity (relative to peak demand) to produce sufficient energy during time periods available wind or solar output is well below average."[15] The technically possible is thus still difficult, needing *expanded transmission, highly flexible demand management*, and lots of *excess capacity*, which is expensive.

Could the US grid system operate with very high levels of variable renewables? In 2012, the US NREL assessed whether a system "largely powered by renewables" was feasible. It found that the then-available technologies were "more than adequate to supply 80% of U.S. electricity generation by 2020."[16] The study examined many scenarios but focused on one in which 80% of generating capacity was provided by renewables, which in this case included nuclear, with 50% of capacity from wind and solar. This mix was sufficient to "balance supply and demand at the hourly level."[17]

But not without lots of work. There has to be investment in grid flexibility and transmission infrastructure to move the electricity from generation sites to the load centers where it is consumed. *Grid flexibility* is central in clean energy discussions; it describes the ways electricity grids need to manage a high reliance on wind and solar. It calls for expanding and improving interoperability among grids; enhancing storage capacity; rapidly sharing reserves among systems; managing demand with time-of-use pricing and other measures; and having natural gas or nuclear backup, because "adequate capacity from dispatchable sources is required to ensure delivery of necessary generation year-round."[18]

The NREL report anticipated that offshore wind, with presumably higher capacity factors than onshore, will play a large role. The storage capacities of compressed solar also contribute. Hydro and geothermal will help, but will add less than others due to resource limits. The study assumed some electromobility, with EV batteries supporting storage. Efficiency gains are essential. The report did not foresee manufacturing or materials barriers to scale-up, though, of course, other experts may disagree.

Much has changed in the last decade. Costs for wind and solar have fallen dramatically, as discussed in Chapter 5. Offshore wind and CSP will become more competitive. Still, the NREL report highlights the challenge of having high levels of wind and solar. Not keeping pace are transmission infrastructure, grid modernization, tools that smooth out demand curves, and battery storage. Again, this underscores the imperative of making all parts of the energy system work together.

Should costs matter? Maybe not, but they do. If energy costs rise too quickly, public support for clean energy suffers. Surveys show that people hold positive views about renewable energy and worry (to varying degrees) about climate change and pollution.[19] But they are sensitive to the costs of a commodity they rely on for multiple purposes daily. There is also, as mentioned earlier, the equity issue: low-income groups spend a higher proportion of their income on energy than high-income ones. For a time, escalating costs took the wind out of the sails of Spain's clean energy progress.[20]

Having a system that uses mostly but not entirely wind and solar is a reliable, cost-effective path. Including even 20–30% of firm generation in the mix enhances reliability and reduces costs. This is why most scenarios incorporate a diverse mix. The question is which sources should be included. The candidates are nuclear, CCS, biomass, reservoir hydro, geothermal, or natural gas with CCS. Current nuclear technologies pose economic and safety issues and are unacceptable in many places. CCS is unproven at scale, and "commercial deployment remains nearly nonexistent."[21] Some biomass is problematic. Large hydro sites are geographically limited and raise ecological issues. Geothermal has geological limits, and advanced technology for expanding it is not yet commercially viable.

There are no perfect solutions. Without a doubt, biomass, hydro, and geothermal will play a role. Nuclear power faces formidable challenges but should not be written off, with new designs on the table. CCS faces economic obstacles, and some decarbonization scenarios rely on carbon removal using bioenergy with CCS (BECCS), also unproven at scale. To get to 100% wind and PV, *all* of four things must happen: transmission capacity has to expand greatly; we need flexible, innovative demand management; wind and PV costs have to fall even further; and there must be a major scale-up in seasonal storage. Jenkins, Luke, and Thernstrom conclude: "Any one of these things may well happen, but it is far less likely all four will be simultaneously achieved."[22]

Boxes 6.2 and 6.3 present two practices supporting a transition. District heating and cooling enhances efficiency and helps in applying renewables. Heat pumps help electrify buildings. Both contribute

Box 6.1: District Heating and Cooling
District heating and cooling takes advantage of co-located buildings (e.g. a university, hospital, or military base). Multiple buildings are heated and cooled from a central source through highly insulated thermal piping networks. This avoids the need for individual building systems, saves space, and cuts costs. The heat is distributed by steam or hot water. District energy cuts emissions with efficiency and by facilitating connections with combined heat and power, conversion of waste to energy, biomass, and geothermal. It saves operating and maintenance costs, due to economies of scale, and is flexible in bringing together diverse heating and cooling sources.

Iceland leads the world in terms of the proportion of citizens served by district heating, at over 90%. Other leaders are Denmark and the Baltic states of Latvia, Estonia, and Lithuania. Those with the most renewables powering their district heating systems are France, Norway, Denmark, and Switzerland. According to a 2016 report from Euroheat & Power, the fastest growth in district heating is occurring in Italy, Switzerland, Norway, China, Sweden, and Austria.[23]

Box 6.2: Heat Pumps
Just as batteries allow us to harness clean generation from renewables, heat pumps replace fossil fuel heating from boilers with efficient, low-cost electricity. Taking advantage of the principle that heat moves toward colder temperatures and lower pressures, heat pumps are an efficient, low-carbon way to heat and cool buildings. They extract heat, even from cold air. Using a fluid refrigerant, they produce heat distributed via water in a building. Heat pumps are up to 30% more efficient than natural gas boilers for heating and less costly to operate, although with higher installation costs. Heat pump technology is used for both air-to-air and ground-to-air (geothermal) heating. Financing is an issue, given the higher up-front costs. As with other options, costs are lower for new buildings than they are for retrofits.[24]

to efficiency and expand the avenues for replacing fossil fuels with electricity.

Integrating High Levels of Renewables

Minimizing the excess generating capacity that has to be built while using high levels of wind and solar to run a cost-effective, reliable grid requires planning, coordination, and money. Three elements, as we have seen, are especially needed: expanded, flexible grids; cost-effective electricity storage; and effective electricity demand management, especially in smoothing out the peaks and valleys.

Enhancing flexibility

Flexibility is a common term in work on integrating renewables into electricity grids. According to the *Oxford Dictionary*, flexibility is "the quality of bending easily without breaking." We want this from any complex system, but especially one as essential as an electricity supply system. We always expected electricity systems to bend without breaking, but the challenge of ensuring this now looms larger.

Ideally, excess afternoon solar energy from, say, the eastern US or eastern Europe could deliver power to western areas, and the latter would return the favor later in the day. Yet that calls for transmission capacities that do not exist. Still, we can expand grids to draw on a more widespread and diverse mix of sources. A healthy grid is like a healthy ecosystem: diversity and variety are a strength.

Clearly, "expanding the electricity transmission grid to transport energy and smooth variability between regions will help mitigate renewable energy supply fluctuations."[25] But trade-offs exist, starting with energy losses. A survey of 142 countries calculated that these losses require extra generation that accounts for nearly 1 billion metric tons of CO_2e each year.[26] These losses and associated emissions were relatively modest in countries like the US (6%) and Germany (5%), although costly. But they came to 19% in India and 16% in Brazil. For countries facing disruption or instability, the losses amounted to 50%. This shows the value of investing in transmission and distribution. Losses are higher at long distances; the loss rate in compact Singapore was only 2%.

Storing electricity

The obvious solution here is chemical battery storage. This is a form of time-shifting: store excess energy when available and draw on it later when needed. The holy grail of clean energy is low-cost, long-term, energy-dense chemical battery storage. There is progress, and battery costs are falling, but battery storage is not yet where it should be. Most storage now occurs not chemically, with batteries, but physically, with water and gravity. Some 95% of storage is pumped hydro; this uses excess energy to pump water up a mountain or other elevation and, using gravity, releases it to drive turbines. Water is efficient for storage, delivering good return on energy invested. If pumped hydro storage is cost-effective and widely used, why not scale it up? The answers are the same as for hydropower: topography and site limits. Pumped hydro also has low energy density. Compressed air can help but does not offer long-term prospects at the needed scale.

Enhanced storage adds to grid flexibility. In generation, it helps avoid supply disruptions and meet peak demand; in transmission, it manages congestion by spreading out electricity flows. It supports distribution by making up for short-term outages and backing up microgrids. A form of distributed generation, microgrids "help deploy more zero emissions energy sources, make use of waste heat, reduce energy lost through transmission lines, help manage power supply and demand, and improve grid resilience to extreme weather."[27] And there is thermal storage. Materials like rocks, cement, and molten salts store heat, as does hydrogen, examined in the next chapter.

Still, looking to mid-century, the key enabling technology is battery storage. Batteries use chemical reactions to store and discharge energy. The current technology is lithium-ion batteries that power EVs and devices like phones and laptops.[28] These have several advantages: low costs at a small scale, high energy density, low maintenance, rechargeability, and high discharge rates. Still, as of 2022, batteries are not yet able to store energy cost-effectively for more than about four hours. To achieve high wind and solar generation and grid integration, the IRENA notes, "electricity will need to be stored over days, weeks, or months."[29] The IRENA projects that this will occur largely with batteries; pumped hydro will fall from a current 95% to under half of storage. Storage from other than pumped hydro would grow from 162 GW in 2017 to 6,000–8,500 GW in 2030 – a growth factor of forty or more. In the IRENA's analysis, by 2030, pumped hydro makes up 45% of storage, EVs are 21%, and 23% is concentrated solar. The rest is in utility-scale storage, electric buses, and other sources. Lithium-ion stationary battery costs may fall 54–61% by 2030, car batteries by even more. Stationary storage has demanding charge and discharge cycles and costs more than mobile storage, adding to the latter's value.

A 2019 Massachusetts Institute of Technology study underscores the issues with battery technology and the marginal costs of moving to a system based entirely on variable renewables.[30] It found that battery storage costs would have to fall by 90%, to about $20 per kWh, to achieve a reliable electricity system. MIT studied two decades of data from four states – Arizona, Hawaii, Texas, and Massachusetts – and found that "storage capacity below $20/kWh could enable cost-effective baseload power" in these states. It also found that "Meeting demand with other sources during 5% of hours can halve electricity costs."[31] Having some firm capacity reduces costs. On a per kWh basis, this requires storage at $150 per kWh rather than $20, a far more achievable goal.

Managing demand

Chapter 4 considered the importance of energy efficiency as an invisible resource. It may be hard to appreciate what we do not use, but not using energy avoids having to produce it. With electricity, it is not just a matter of not using it but of *when* it is not used. The visual of a daily load curve (representing the ups and downs in demand) illustrates the task facing electricity systems: unable to store electricity on a large scale, they need enough generating capacity on hand to meet the *peak loads*. This is not just a daily challenge; weather and seasonal changes cause long-term fluctuations as well. Historically, grids relied on natural gas peakers or hydro to meet demand spikes. Gas peakers ramp up and down quickly, although at a high marginal cost. And, of course, they use fossil fuels.

Demand management involves lowering the peaks in loads so there is less of a need for stand-by generation. Many options exist. One is working with end users to cut consumption, such as through residential or business energy audits. The core of it, however, is time-shifting: changing daily and long-term load patterns with tools like dynamic pricing, where costs are higher at demand peaks and lower in valleys.[32] Technology lets us program appliances to run at midnight rather than in the late afternoon, when retail prices are higher. Two-way information flows enable smart grids to deliver real-time data on prices so consumers can choose to save money while smoothing demand. Better overall efficiency is part of managing demand, but the more specific goal is engaging in the practice of "peak load shaving" in managing demand and the time-shifting of daily and seasonal electricity loads.[33]

The Smart Grid: Technology Helps

The adjective *smart* is all around us; we speak of smart phones, cities, and TVs. All rely on information and communications technologies at the heart of a smart grid. The Office of Electricity at the US Department of Energy sees the smart grid as "the digital technology that allows for two-way communication between the utility and its customers, and the sensing along with the transmission lines is what makes the grid smart."[34] Technologies "work with the electrical grid to respond digitally to our quickly changing energy demand."[35] For Heinberg and Fridley, a smart grid is made up of "integrated communications, sensing and measurement devices (smart meters and high-speed sensors deployed throughout the transmission network), devices to signal the current state of the grid, and better management and forecasting software."[36] Digitization is a theme; smart grids build on advanced sensing, monitoring, and real-time data.[37]

A smarter grid promotes efficient transmission and consumption while using monitoring and sensing that provides actionable feedback to users. It helps integrate renewables by optimizing renewables, forecasting resources, and integrating EV batteries in grids. Two-way communication allows for an *intelligent integration* of distributed generation from PV, wind, and, someday, maybe even small, modular nuclear reactors.[38] A smart grid reduces electricity demand peaks that increase costs and require extra capacity. ICT promotes security and resilience by giving grid managers real-time data and analytic tools. Table 6.1 lists the benefits of a smart grid.

A 2011 analysis by the Electric Power Research Institute examined the costs and benefits of a smart US grid over a twenty-year period. It estimated the costs at $338 to $476 billion ($17 to $24 billion a year), and the benefits at $1.3 to $2 trillion ($65 to $100 billion

Table 6.1: Benefits of a Smart Grid

Supply Side	• Optimizing the facility's utilization and reducing the need for excess capacity that peak-load power plants provide • Improving the connection between and operations of generators of all sizes and technologies • Reducing the entire electricity supply system's environmental impact	
Electricity Network	• Preventive maintenance and remote grid management through better monitoring and control features • Minimizing energy losses through efficient energy routing • Increasing the degree of automation and "self-healing" responses to system disturbances • Incorporating distributed energy resources (DERs) and PHEVs effectively	Increase or maintain the • Reliability • Security • Power quality • Resilience • Energy and economic efficiency • Environmental sustainability of the energy system
Demand Side	• Providing consumers with better information • Increased responsiveness and demand flexibility • Enhanced efficiency through better management options and greater awareness of energy consumption • Giving power consumers a more participative and active role • Enabling innovative services and applications	

Source: Johann J. Kranz and Arnold Picot, "Toward an End-to-End Smart Grid: Overcoming Bottlenecks to Facilitate Competition and Innovation in Smart Grids," National Regulatory Research Institute, 2011.

each year). Every dollar invested returned between three and six dollars in benefits, due to more efficient operations, better asset use, enhanced security and resilience, more flexible generation and storage, fewer outages, and the ability to integrate high volumes of variable renewables.[39]

Smart grids offer money-saving opportunities that often go unrealized. Some of the reasons for this are similar to those explaining the underinvestment in efficiency: too little information and expertise, lack of capital, short-term horizons, organizational inertia, and other priorities. With the smart grid, however, there is another explanation: it exhibits the characteristics of a *public good*. Public goods are ones that "do not dwindle in supply as more people use them" (non-rival) and that are available to everyone (non-excludable).[40] National defense is a classic illustration. The opposite is *private goods*, which are rival and excludable. Electricity has aspects of a private good: my consumption makes it unavailable to others. At the same time, a reliable, secure electricity system also is a public good. Investments support a reliable grid for all users; my use of it does not exclude others who also benefit from using it.

The movement toward a smart, or at least much smarter, grid is being driven by rapid changes in the electricity sector. These include wind and solar expansion, the growth of distributed energy, and the need for storage.[41] Opportunities lie in falling technology costs and such analytic tools as artificial intelligence, blockchain technology, and methods for optimizing grids.[42] Yet the smart grid is more than technology. Consider *collaborative smart grids*, which use ICT to create "new business models based on engagement, behavioral change, and benefits distribution."[43] Old grids were centralized, unidirectional, and hierarchical; new grids are distributed, bidirectional, networked, participatory, and resilient. Or at least they will need to be if the clean energy transition is to succeed.

There are two ways in which an ICT-enabled grid is collaborative. One is its reconceptualizing of products as services, less as energy or cars and more as mobility. For example, a smart grid gives short-term car users data about optimum use times to smooth out demand and maximize renewables use. Second is its creation of "communities of consumers" to collaborate on energy sharing, use, and storage.[44] People share storage and collaborate in energy production and use. Many experts see collaborative grids as a means of linking clean energy with social change – as enhancing *energy democracy* (Chapter 9).

With their falling costs, the challenge of variable renewables turns on how we integrate them into grids while electrifying more final energy use. Modern grids have to cope with many moving parts: variable, distributed generation; two-way information and energy flow; efficiency demands; EV growth; new technologies and analytical tools. This underscores the need to think of energy as a complex system where a change in one part cannot be isolated from what happens elsewhere.

Electric Vehicles and the Clean Energy Transition

Some 95% of global mobility is fueled by oil. The internal combustion engine is a poster child for carbon lock-in. Recent trends show that carbon is locked even more into mobility than it is into electricity generation. Globally, 60% of electricity relies on fossil fuels, but nearly all transport runs on petroleum. If transport emissions do not fall big time, decarbonizing is a pipe dream. One in four tons of energy-related emissions are from transport, 60% of those from light-duty vehicles, mostly passenger cars. My focus here is cars and light trucks, although with some attention given to other forms of transport.

The basics of electric vehicles

EVs come in four categories. Battery electric vehicles (BEVs) are an alternative technology and a complete departure from ICEs.[45] They are the cars and possibly the trucks of the future. BEVs run entirely on batteries and require access to charging stations. In contrast, hybrid-electric and plug-in hybrid-electric vehicles (HEVs and PHEVs) combine an ICE with battery technology. HEVs are designed to recharge while in use and draw on both battery power and a gas engine. PHEVs are the same, but they allow for battery recharging. A fourth category is fuel cell electric vehicles (FCEVs), which use compressed hydrogen to drive electric motors; they will be discussed separately.

HEVs and PHEVs are a gain over the standard ICE for clean energy, but they still directly use fossil fuels and emit carbon. However, they play a role in the scenarios. In percentage terms, EV sales are rising rapidly. Light-duty sales in the first quarter of 2021 were up 140% over that same period of 2020. The hands-down leader in market share was Norway, where EVs were 75% of new sales. Others with high sales shares were Iceland (45%) and Sweden (32%). The big markets had lower rates – 6% of sales in China and just over 2% in the US. The world share was 4.2%, well above the 2.5% in 2019.[46] Figure 6.4 shows changes in electric vehicle stocks in the last decade.[47]

million

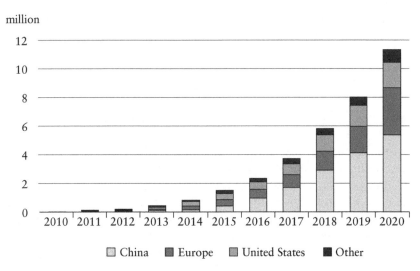

Figure 6.4: Global Electric Vehicle Stock by Region, 2010–2020
Source: IEA, *Global EV Outlook 2021*, https://www.iea.org/reports/
global-ev-outlook-2021?mode=overview.

Many countries aim to speed up the clean energy transition by
banning sales of gas and diesel vehicles or restricting their use.[48]
Reflecting its aggressive promotion of EVs, Norway requires that all
new cars and vans be zero-emission by 2025. Iceland, Ireland, and
the Netherlands are banning sales of fossil-fueled light-duty vehicles
starting in 2030; Scotland and the UK will be phasing out gas and
diesel vehicles in the 2030s. Spain and France want to allow only EV
sales as of 2040. Locally, many European cities have already acted or
have plans to create car-free centers, charge congestion fees, or ban
gas and diesel vehicles from their centers; these include Oslo, Paris,
Amsterdam, and Brussels.

Manufacturers are getting the signal. Almost all the major
automakers plan to expand their EV fleets. As one expert put it,
"Established carmakers around the world are ripping up their
business models in the hope of adapting to a new world in which
electricity replaces gasoline and diesel."[49] In 2019, Volkswagen
announced a plan to spend $34 billion over five years to produce
an electric or hybrid version of all its cars. It wants four of every
ten sales to be electric by 2030. Other leaders are Renault-Nissan-
Mitsubishi, Tesla, and China's Geeley, which owns Volvo. Although

Tesla led sales in the early 2020s, Volkswagen and other giants plan to use their large manufacturing base to become EV leaders. The news from GM in 2021 was that it would sell only zero-emission cars and trucks by 2035, months after California announced a 2035 plan to allow only zero-emission new car sales in that state.[50]

Prospects for electric vehicles

Decarbonization scenarios place big bets on electrified mobility. The IEA's Stated Policies Scenario (what happens if all existing, stated policies are implemented) projects that 145 million light-duty EVs will be on the road by 2030, 7% of the global fleet. Its more ambitious Sustainable Development Scenario (what needs to happen for minimal carbon by mid-century) calls for a light-duty EV fleet of 230 million (13% of the total).[51] The IEA projects rapid growth in electrified two-wheelers, buses, and truck transport. Bus fleets grow to 3.6 million in the Stated Policies Scenario (10%), and to 5.5 million in the Sustainable Development Scenario (15%). Recognizing the barriers to electric trucks, especially long-haul freight, it projects EVs in this area of only 1–3% under the two scenarios.

What are the barriers for EVs generally? One is cost, driven by battery technology. Without tax supports and other subsidies, EVs still cost more than ICEs, although prices are falling. Fuel and maintenance costs are low for EVs, but sticker prices are a barrier. Another issue is *range anxiety*. Although EV ranges are increasing and charging stations expanding, psychological and practical barriers remain. A further worry is access to critical materials, including cobalt, lithium, copper, and nickel.

Cost is a major challenge, but a receding one. Batteries account for most EV costs. They will fall with economies of scale and technology gains. The issue is battery energy density, particularly mass density, the "amount of energy contained per unit of mass of an energy resource."[52] Liquid fuels historically won out for transport due to their high mass density. Batteries require mass that limits range and increases costs. In turn, higher density lowers costs and expands range. Economies of scale help. As production volumes grow, the cost of each additional unit (the marginal cost) falls. This explains much of the falling cost of wind and PV and applies to vehicle batteries as well.

In sum, the more chemical batteries manufactured, the lower the cost. As density improves, costs fall, batteries hold more power, and ranges increase in a virtuous cycle. As the IEA projects in its *Global EV Outlook*, lower costs will come with better chemistry, higher density, less cobalt, larger plants with increasing economies of scale, and enhanced battery capacity. Bloomberg New Energy Finance reports that battery prices fell 89% in real terms from 2010 to 2020 to an average of $137/kWh.[53] The costs of EVs are expected to be level with those for ICE vehicles by 2024 as battery costs fall below $100 per kWh.[54] In September 2021, for the first time, the global new EV sales share passed 10%.[55] Probably few numbers in this book will change as rapidly as those on EV sales and costs.

Tesla demonstrates the value of learning over time and cutting costs as production expands. Its EV success "is based on the economies of scale ... which allow them to be manufactured significantly cheaper as we manufacture more and more."[56] As Tesla makes more cars, experience and volume reduce the costs; "with each model and each factory, the learning process improves."[57] Once companies like GM and Volkswagen enter the picture, these economies of scale will be magnified, and costs should fall even more.

Critical materials are an issue. Cobalt and lithium are particular concerns; both are essential to current chemical battery designs. Lithium is found in South America and Australia; the growing demand will have to be met by sources in Chile, Argentina, Brazil, and Australia. Cobalt is a bigger worry. Two facts are telling: 60% of global cobalt supplies come from the Democratic Republic of Congo (DRC), and 90% of global refining capacity is located in China. The DRC is poor, with endemic corruption; the *Democracy Index* rates it one of the most authoritarian countries in the world, near North Korea.[58] Beyond politics, concerns include dangerous work conditions and child labor. Cobalt supplies exist in Australia, Canada, the Philippines, and South Africa, but DRC's "dominance presents a growing dilemma for carmakers and others in the supply chain as they look to meet a rapid increase in demand for electric vehicles and batteries."[59] China dominates refining, but capacity exists in Finland, Canada, and Norway. Reducing cobalt needs is a priority for EVs.

EVs replace fossil fuels with clean electricity. Hundreds of millions of EVs expand storage. Vehicles are used on average less than 5% of the time, making batteries part of the storage solution: EVs can draw electricity from the grid to charge batteries at low demand and send it back to the grid at peaks. Vehicle-to-grid (V2G) and vehicle-to-home (V2H) technology adds grid storage capacity.[60]

Are electric vehicles that much better?

EVs do not eliminate transport energy use or emissions. Movement requires energy. Yet "driving on electricity produces significantly fewer emissions than using gasoline and is getting better over time."[61] How much better depends on how the electricity is generated. A few years ago, the average EV in the US could travel 88 miles on the equivalent of a gallon of gas. This was only 10% better than EVs that were tested a few years earlier. But there were differences across regions, depending on the electricity system. The best results were Upstate New York, which uses hydro, where EVs moved 231 miles per gallon of gas equivalent. In California, which is expanding wind and solar, it was 122 miles. In Texas, having more fossil fuels, it was 68 miles. A recent study found that EVs require more energy to manufacture but less to use, especially with clean electricity. After about 21,000 miles of use, EVs are far more carbon-efficient.[62]

These numbers will only get better. Expanded use of gas and renewables relative to coal means that electricity is cleaner. Gains in EV technology and efficiency help, with the Tesla 3 now considered "one of the most fuel-efficient vehicles on the market."[63] A Tesla 3 in a relatively clean system like California's got 161 miles per gallon of gas-equivalent, one-fifth of the gas requirement for a new ICE vehicle. Studies find that even large electric SUVs perform far better than their ICE counterparts.

Again, the lesson is that the various parts of the energy system must be coordinated. Having fleets of EVs charged by electricity from coal plants is not progress. An unsmart grid may accommodate some wind and solar but quickly reaches an upper limit. EVs increase the burdens on a smart grid but help smooth out demand. Renewables, electrification, and efficiency have to be coordinated – a role for public policy.

The Harder Stuff: Decarbonizing When Electricity Is Not an Option

Electricity can replace many liquid, carbon-based fuels. For others, the low energy intensity of chemical batteries is an obstacle. Long distances, large loads, and limited charging options make electricity poorly suited to heavy-duty, distance transport. Just as problematic are fossil fuels used to produce industry heat and as feedstocks. Distance trucking, aviation, ocean shipping, and heavy industry will be the toughest sectors to decarbonize.

Freight, aviation, and ocean shipping

Getting carbon out of passenger travel by mid-century will be difficult, but there is a path. Other forms of transport are more challenging. Road freight, aviation, and ocean shipping accounted in 2019 for some 10% of global CO_2 emissions; all are projected to grow. Aviation demand will grow at least 50% and maybe more by 2040. Although ocean shipping is more carbon-efficient than road transport, most global trade relies on shipping that accounts for up to 3% of CO_2. Oil demand for long-distance trucking may increase by 40% by 2040.[64]

For such transport, oil density is even more valuable than in passenger vehicles. Samantha Gross writes that "Replacing oil will be easier in small vehicles carrying lighter loads with frequent opportunities for refueling."[65] Likewise, "large, heavy vehicles such as trucks, tractors, and cargo ships require batteries too heavy to be practical in most instances, particularly if they are traveling long distances."[66] With current technologies, battery weight is an obstacle to getting airplanes off the ground.

For small trucks using regular routes and making frequent stops for urban deliveries, small battery size and recharging options make electricity feasible. For trucks going long distances, with heavy loads and high power demands, batteries are less workable. At this scale, batteries are prohibitively heavy, reduce hauling space, and require excessively long charging or upgraded grid infrastructure for fast charging. For now, the options for large trucks are low-carbon biofuels or, eventually, green hydrogen.

Aviation and ocean shipping will be among the last sectors to decarbonize. Both can start with better fuel efficiency, where aviation especially is progressing. Batteries may someday be feasible for smaller aircraft on short flights. In the meantime, the options are low-carbon biofuels and, well into the future, hydrogen for fuel cells or used directly as a jet fuel. Even biofuels, a low- but not zero-carbon option, face barriers: limits in feedstock supplies, issues with fuel conversion technologies, and high costs. Another issue is that biofuels "are strongly linked to land use changes" that cause environmental harm, remove land from food production, and promote deforestation that offsets their carbon benefits.[67]

On the water, as on land, transport involving short trips with recharging opportunities, such as ferries, are best suited to electrification. Ocean shipping, with its long distances and large cargos, is poorly suited to electrification with the current technology. In 2018, the OECD's International Transport Forum thought that ocean shipping could "almost completely decarbonize" by 2035, cutting emissions 80% below projections.[68] This could happen with the use of biofuels, fuel cells, hydrogen propulsion, and improved hull designs and surfaces. Although this may be technically possible, the gap between the possible and likely here is wide. Still, there is a path to at least greatly cutting emissions by 2050.

Aviation illustrates the multiple motivations for clean energy innovation. Electric solutions eliminate emissions and can cut fuel costs by up to 90%, reduce repair costs by half, and eliminate most noise. Half of all flights are 800 kilometers or less, and many could be electrified in the next few decades. Airbus says it wants 100-passenger electric planes ready by 2030.[69]

Heavy industry: a big lift

About a third of world emissions are from industry. In light industry – food processing and textiles – electricity is an option. Aggressive energy efficiency improvements, technology production innovation, and someday green hydrogen are solutions. Heavy industry sectors are more difficult. Some 90% of industry emissions come from a dozen sectors, including the top three of iron and steel, chemicals, and cement. Other large emitters are aluminum, oil refining, and pulp and paper. Emissions come from on-site combustion of fossil fuels, manufacturing processes, and indirect electricity and heat. The mature economies accounted for most industrial emissions until a few decades ago, when places like China began their explosive growth. Today, the carbon-intensive growth is occurring in Asia and will account for most of the increases in industry emissions in this century. Decarbonizing heavy industry is tough "due to their relatively high share of emissions from feedstocks and high-temperature heat compared to other sectors."[70] From 1990 to 2014, power, buildings and transport emissions grew an average 0.9% a year, industry at 2.2% a year.[71]

These sectors *could* get to net-zero carbon by 2070.[72] Much would have to happen quickly, and it would have to be well coordinated. A transition would start, of course, with efficient energy use, both at a process level (more efficient electric motors, minimizing distribution system losses, managing leaks) and at a more integrated systems level (redesigning processes or products).

Electrification can meet needs not requiring high levels of heat, such as for general heating. Hydrogen has promise, although it must be green and produced through electrolysis, which, as Chapter 7 explains, currently only accounts for about 5% of production. Hydrogen can be used directly as a heat source with fuel cells, or to produce alternative fuels like ammonia, methane, or methanol. The fallback is CCS, perhaps utilizing the carbon in synthetic fuels, plastics, or building materials. The economics of carbon capture are examined in Chapter 7. Finally, we can expect long-term gains from redesigned production processes and products.

Access to Electricity, Sustainable Energy, and Development

In parts of the world, electricity is not available. This is a barrier to development. Energy's role in quality of life is reflected in the UN's Sustainable Development Goal 7 (SDG7), which is to "ensure access to affordable, reliable, sustainable, and modern energy for all" by 2030.[73]

Most of us are used to having reliable electricity. This is not the case in poor countries, where electricity is unavailable or unreliable, especially in rural areas. But there has been progress: the World Bank and IEA reported in 2020 that the global population without access to electricity had fallen from 1.2 billion in 2010 (17% of the global population) to 789 million (10%) in 2018. Bangladesh, Kenya, and Uganda gained the most.[74] Access is especially low in countries with extreme poverty and large rural populations: Niger, Chad, Burkina Faso. In numbers, deficits are highest in India, Nigeria, and the DRC. One effect of the 2020 pandemic was to slow progress in expanding access, but the trend should resume.[75]

In middle- and high-income countries, electrification is all about decarbonizing the energy system. They have affordability issues, to be sure, but access is taken as given. In poor countries, electrification is about going from no access to reliable access. Electricity is a ticket to a better life. Yet deficits are common in poor areas. The worst deficits are in Sub-Saharan Africa, where, even in 2018, nearly half the population lacked access. Some countries do better than others. Urbanization helps; global access in urban areas is 97% but 85% in rural areas. India is making progress, with over 95% of the population having access, but this still excludes more than 50 million people. Low access is common in land-locked countries (e.g. Rwanda) and those with inadequate grid systems (e.g. Somalia).

Wind and PV offer potential for expanding access. With electricity, the challenges come in the last mile of bulk grid connection. Wind and PV deliver on a *distributed* as well as *utility* scale. Distributed energy generation is "the production of electricity near its point of use."[76] It may be connected to bulk grids or can operate independently. Off-grid distributed generation does not require a bulk infrastructure and can be organized at a household, community, or business scale.

Distributed generation through PV, small-scale wind, or beneficial biomass promotes sustainable development. It enhances economic well-being by enabling computing, lighting, communications, and other activity. It promotes social well-being with lighting for public safety and health care, and lighting and cooling for education. It allows people to replace harmful indoor cooking using heat from biomass with electricity. Water sanitation and refrigeration become feasible. And generating with PV and other clean sources reduces climate- and health-related pollution.

India illustrates the merits of distributed PV generation.[77] The government is committing to solar in order to move away from coal. Distributed generation electrifies while minimizing investments in infrastructure and transmission. It should be gaining more support from the rich countries.

Electrify Everything, But How?

Electrification is an essential pillar of the clean energy transition, and the main area in which to expand the use of renewables. Transport depends on fossil fuels that green electricity, green hydrogen, and advanced biofuels eventually will replace. But electricity has to do most of the work. At the same time, those without electricity have to be plugged in. Until the pandemic, there was steady progress, as is clear from the SDG7 reporting on energy. That was interrupted, but it is not over.

In summary, the world must use a great deal more electricity by mid-century. The falling costs of renewables, growing investment, and technology innovation are keys to cleaner electricity. Yet considerable *systems innovation* is needed: smarter grids; flexible generation; transmission capacity; EVs as part of storage; smoothing out electricity loads; and other innovations. All of this underscores the importance of capable governance in an energy transition. Not only does the old equilibrium have to be punctuated, governments and related institutions have to be able to connect the dots in guiding, facilitating, funding, and generally promoting the changes that must occur.

The next chapter takes up two lively issues: nuclear power and CCS. Nuclear has long been a source of worry in relation to safety, waste, and security. CCS maintains fossil fuels, releases health-harming pollutants, and emits residual CO2. Still, both meet energy needs and enjoy political support. Many experts see them as essential to a renewables-powered grid. Others think we can do without them. The next chapter examines both, along with the promising option of green hydrogen.

Guide to Further Reading

Gretchen Bakke, *The Grid: The Fraying Wires Between Americans and Our Energy Future*, Bloomsbury, 2016.

Kathryn Cleary and Karen Palmer, "Integrating Renewable Energy Resources Into the Grid," Resources for the Future, April 15, 2020. Part of the *Future of Power Explainer Series* from RFF, which is an excellent resource on electrification basics.

International Renewable Energy Agency, *Renewable Energy Policies for Cities: Transport*, May 2021.

Jesse D. Jenkins, Max Luke, and Samuel Thernstrom, "Getting to Zero Carbon Power Emissions in the Electric Power Sector," *Joule* 2 (2018), 2498–510.

7

Hard Choices and an Opportunity
Nuclear, Carbon Capture, and Green Hydrogen

That wind, solar, and other renewables are even part of the discussion reflects the gains of the last two decades. As recently as 2000, it would have been hard to imagine the combination of technology innovation and economic production that made options like PV, concentrated solar, and wind energy realistic for large-scale development. With help from geothermal, biomass, and eventually ocean sources, these technologies will be the workhorses of the clean energy transition. Yet relying only on variable renewables involves challenges, as discussed in the last few chapters.

It is one thing for the state of Hawaii or the city of Washington, DC, to aim for electricity based totally on renewables. They can draw on other sources and systems for help when needed. It is quite another for an entire country or large numbers of them to do the same thing. It could happen in the timetable envisioned in the scenarios, but not without risks. What if the political power of fossil fuel sectors cannot be overcome and policies are not changed sufficiently? What if investments in smart grids and transmission lag? Will the critical materials be there when needed? Will short-term costs provoke opposition and undermine progress?

This chapter examines three technologies that allow us to hedge our bets: nuclear power, carbon capture and storage, and green hydrogen. The first two are controversial, especially nuclear. What was exalted as the solution to global energy needs in the 1950s now is *persona non grata* in much of the world. CCS is hailed as a way to redefine fossil fuels as clean energy. Green hydrogen is less controversial, but it faces technical and economic challenges, as do the first two.

Although the political controversy varies with each of these technologies, the economic and practical challenges are similar. All three will play a role in the global energy system of the future. Just how big a role depends on the dynamics of clean energy choices (including agreement on what clean energy means), commitment to investing in research and development, and progress in technologies and associated infrastructure. This chapter examines each in turn, considering both their state of development and their mid-century prospects. Later in the chapter Table 7.2 provides a summary assessment.

What About Nuclear Energy?

The technology for using nuclear energy to generate electricity emerged from World War II research in the US to develop the atomic bomb. This is one of many illustrations of how innovative technologies were invented or scaled up during conflicts.[1] The roots of nuclear technology and its role in the early years of the Cold War had a major effect on its development. Nuclear technology was adapted for submarines, and their light-water reactor (LWR) designs became the model for civilian-commercial reactors to generate electricity. Eager to show the civilian as well as military value of the industry, the US committed further to developing and disseminating LWR technology.

Nuclear is often used as a cautionary tale for technologies generally. It was hailed in its early years as a low-cost, reliable, low-polluting solution to the growth demands of postwar economies. It also has become an example of *technology lock-in* (carbon not being the only form of lock-in), where government, investors, utilities, and others bet on a solution, even though alternatives exist. As Robin Cowan notes, the US electricity and defense industries committed to LWRs over heavy-water and gas graphite reactors and promoted them elsewhere. By the mid-1960s, LWR was the technology of choice not only in the US but in most of Europe.[2]

Nuclear energy is a cautionary tale because it has not lived up to expectations, a story that renewables supporters fear could be repeated. Indeed, prominent PV advocate Varun Sivaram worries that, as economic and political interests organize around existing silicon technologies, they will be locked in, just as light-water reactors were for nuclear.[3] Technologies other than ones based on silicon may be PV's future, but any lock-in of the existing technology puts that future in jeopardy.

Nuclear has long been a flashpoint for environmentalists in Europe. Indeed, opposition is strong in many European green parties.[4] In the US, nuclear skepticism is more pragmatic, although anti-nuclear sentiment comes through in the Green New Deal and other proposals. Still, the US now has the largest nuclear fleet in the world – ninety-four reactors producing nearly 20% of its electricity. Yet its future is in doubt. The Nuclear Regulatory Commission (NRC), which licenses plants and oversees safety, has extended many plant licenses; even so most may shut down by the late 2030s.[5]

Nuclear energy poses a dilemma for clean energy advocates. Like renewables, once plants are operating, they deliver relatively clean (low-carbon, minimal pollution) electricity. They are reliable sources of baseload power, with capacity factors of 80–90%. At the same time, they use a technology that in theory, and at times in practice, may have catastrophic consequences; Chernobyl and Fukushima are the leading examples. They generate radioactive wastes that are hard to dispose of and last for centuries. Like other thermal technologies, nuclear plants withdraw lots of water in generating steam to drive turbines, and this may or not be cooled sufficiently before it is returned to its source. Nuclear plants use uranium, raising fears of nuclear proliferation and opportunities for terrorists. Finally, warmer temperatures have forced plants to shut down for short periods when water is too warm to use as a coolant.[6] Existing plants face other climate risks, like droughts and sea-level rise: "the relationship between nuclear power and climate change is a two-way street."[7]

Politics is an issue, especially in Europe. Yet the biggest obstacle is economics. This is related to safety; capital and operating costs are high partly due to strict licensing and oversight. The industry is affected by events like Fukushima, Chernobyl, and Three Mile Island, when public fears surface and are validated.

The global nuclear industry

The high hopes of the 1950s for nuclear energy have not been realized. Nor have they been totally unfulfilled. In 2020, just over 10% of global electricity was generated by nuclear energy, making it the second source of low-carbon power, behind hydro. As of March 2020, according to the World Nuclear Association (WNA), 440 reactors operated in thirty countries, with fifty more under construction.[8] The US led the world in total nuclear generation, followed by France, China, Russia, and South Korea. The US generated twice as much nuclear electricity as France and over three times that of China. With three-fourths of its electricity coming from nuclear, France easily led the world in nuclear *share*, although Hungary, Slovakia, and Ukraine get over one-half of their electricity from it, and Sweden, Belgium, Bulgaria, Switzerland, Slovenia, and Finland at least one-third. Table 7.1 lists existing nuclear fleets, their share of electricity, and plants under construction in several countries.

Before Fukushima, Japan generated one-fourth of its electricity from thirty-three plants, and it aims to return to that level. As of the start of 2020, only nine plants were back online and seventeen were in the approval process for restarting.[9] India derives only 3% of its power from its twenty-two plants but is expanding, with seven plants being built. China has ambitious expansion plans. Its current

Table 7.1: Nuclear Fleets, Generation Share, and Reactors Under Construction, 2019

Country	Number of Reactors	Share of Generation (%)	Reactors Under Construction
Argentina	3	5.9	1
Brazil	2	2.7	1
Canada	19	14.9	0
China	48	4.9	11
France	56	70.6	1
Germany	6	12.4	0
India	22	3.2	7
Japan	33	7.5	2
Mexico	2	4.5	0
Pakistan	5	6.6	2
Russia	38	19.7	4
South Korea	24	26.2	4
Sweden	7	34.0	0
Ukraine	15	53.9	2
United Kingdom	15	15.6	2
United States	95	19.7	2

Source: World Nuclear Association, *World Nuclear Performance Report*, 2019.

forty-seven reactors meet 4% of its huge electricity needs, but China only accounts for one-fifth (11 of 53) of all reactors being built globally. Bangladesh, Belarus, Turkey, and United Arab Emirates are building their first plants. Other countries with expansion plans include South Korea, Russia, and the United Kingdom.

Not all countries are committed to expanding their nuclear capabilities. France aims to diversify by reducing its share to about one-half of electricity by 2025, although it is building a new reactor. But Germany is the big skeptic. It has closed eleven plants since 2011 and will close its six others (producing 12% of electricity) by 2022. The gap is filled mostly by coal, causing Germany to miss its target of cutting emissions below 1990 levels by the end of 2020. The phase-out reflects German public opinion and the Green Party's influence: "Resistance against nuclear power is the party's DNA."[10] Long controversial, the nail in the coffin for Germany's nuclear plants was Fukushima in 2011.

A study of the effects of nuclear shutdown in Germany illustrates the hard choices. Because nuclear was replaced largely by coal, Germany's CO_2 emissions rose 5%, some 36 million tons. This increased particulate and other emissions. According to one study, the excess pollution "likely killed an additional 1,100 people per year from respiratory or cardiovascular illnesses," and imposed health and mortality costs of $12 billion a year.[11] This is more than the nuclear plants would have cost to maintain, even after accounting for the required safety and waste disposal expenses.[12]

The effects of nuclear power on historical global emissions are illustrated in Figure 7.1. In the last half-century, nuclear and hydro were "the backbone of low-carbon electricity generation" globally. Nuclear avoided 55 billion tons of CO_2, nearly two years of energy-related emissions.[13] Yet existing fleets are aging; the average age of plants is forty years in the US and a bit less in Europe. Without a new generation of reactors, experts fear a "nuclear fade" in which shuttered plants are replaced by fossil fuels while renewables struggle to keep up.

In the US, safety issues matter; larger barriers are capital and operating costs. Fukushima forced an assessment of the ninety-five US plants then operating as well as the adoption of new safety standards. Based on a need for low-carbon electricity, the Nuclear Regulatory Commission extended the licenses of existing plants. The *Carbon Brief* reported in 2018 that five US plants had closed in the

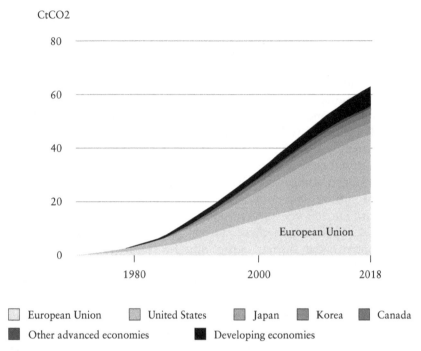

Figure 7.1: Cumulative CO2 Emissions Avoided by Global Nuclear Power, 1971–2018
Source: IEA, *Nuclear Power in a Clean Energy System* (2019), https:// www.iea.org/reports/nuclear-power-in-a-clean-energy-system.

last five years; nine more are likely to close in the early 2020s.[14] Six plants set to close remained open due to policy action in New York, New Jersey, Illinois, and Connecticut.

The US illustrates the challenges facing the industry. As of November 2021, it had ninety-three reactors operating in thirty states, with two under construction.[15] Almost all the plants were built in nuclear's golden years of 1967 to 1990. Operating costs fell from an average $42 per MW in 2012 to $30 per MW in 2019, and capacity factors averaged 94% in 2019, up from 70% in 1990. Despite their operating efficiency, plants have trouble competing against cheap natural gas and wind subsidized by such incentives as the Production Tax Credit (see Chapter 8). Adding to the challenge are the deregulated electricity markets in many states, where utilities cannot absorb the capital costs of new plants into their rate base.

The industry view is that electricity deregulation, cheap natural gas, and subsidized wind have undermined the economics. A carbon tax would help by recognizing the avoided social costs of low-carbon electricity.

With the world's largest fleet, the US faces hard choices, as do many other countries. The 100% scenarios reject nuclear and place their bets entirely on wind, solar, and other renewables. Still, most experts doubt that a clean energy transition and decarbonization can occur without nuclear.

The Future of Nuclear Energy

It helps to separate nuclear's past from its future. One expert wrote in 2018 that Germany's "almost religious antinuclear attitude doesn't leave room for advances in technology."[16] Advances in technology and design – going smaller, safer, and at less cost – deserve attention.

Two things have to happen if nuclear is to have a major role: costs need to come down and confidence in safety has to improve. If neither happens, or even only one, its future is uncertain, at least without generous financial support. The issue in many places is its capacity to compete in the marketplace. Advanced nuclear (third-/fourth-generation) designs differ from the large light-water reactors of the past. Advanced nuclear is smaller; relies on modular design and construction, which cuts capital costs; uses passive safety to enable default shutdowns; recycles spent fuel, which minimizes waste disposal; and operates for long periods with minimal maintenance.[17]

Nuclear's future lies in these advanced reactors – new designs that are small, standardized, and use modular construction, leading to lower costs and presumably more streamlined licensing. They incorporate passive safety systems that require "no active controls or operational intervention to avoid accidents in the event of malfunction, and may rely on gravity, natural convection, or resistance to high temperatures."[18] Small reactors (300 MW or less) with passive safety rely on simpler designs, use fuel more efficiently, produce less waste, have longer lives with lower maintenance costs, and may be combined for uses at flexible scales.

These are lots of promises, but the trends are encouraging. A company called Nuscale leads in small, modular reactor design and has received a "preliminary design approval" from the US Nuclear

Regulatory Commission. It is building a complex of small reactors to generate electricity in Utah by decade's end.[19] A partnership of Bill Gates and GE Hitachi Nuclear Energy is advancing a design (Natrium) using liquid sodium as a coolant with a capacity to store energy, enabling it to quickly ramp up and deliver electricity in a renewables-driven system.[20] Such designs help not only in integrating wind and solar, but in producing green hydrogen, meeting industry heat needs, and generating electricity at remote sites.

Advanced nuclear technologies could help with tough decarbonization challenges, but are they worth the investment in research and deployment that is needed? My view is that no feasible options should be eliminated. Nuclear is controversial, expensive (at least in capital costs), risky (a source of controversy), and generates radioactive waste that remains with us for a very long time. On the other hand, once plants are operating, they deliver reliable, low-cost baseload electricity, with no CO_2 or local air pollutants. Countries are investing in nuclear, and it features in many decarbonization scenarios. Its future may be brighter than its past.

Whatever the future of nuclear, it helps to focus on the potential of new technologies with a realistic chance of scaling up and not just those of the past. As Ben Soltoff writes in *Greenbiz*, "You're now buying a model 2020 Mercedes. You're not buying a Peugeot from 1965."[21]

Carbon Capture and Storage

There are ways to use fossil fuels while generating minimal carbon. Technologies exist for extracting CO_2 from emission streams and injecting it below ground. Carbon capture and storage removes nearly but not quite all the CO_2 from emission streams. Yet it is costly, still causes environmental damage, and, of course, it perpetuates carbon lock-in.[22] For some people, it is just another excuse not to decarbonize. For others, it is a vital tool of the clean energy transition.

The IPCC defines CCS as "a process of the separation of CO_2 from industrial and energy-related sources, transport to a storage location and long-term isolation from the atmosphere."[23] The carbon must be sequestered indefinitely in old oil and gas reservoirs, deep saline aquifers, oceans (the water column or ocean floor), or transformed into carbonates.[24] If the carbon is turned into products, such as biofuels or bioplastics, the process is CCUS (*utilization* and storage). Good processes claim to capture 90% or more of the CO_2. For critics, CCS technologies are a cynical case of eating your cake while still having it – of decarbonizing with carbon. Still, few scenarios envision net zero by mid-century without at least some carbon-based electricity and mobility, so there are reasons for including CCS on the clean energy agenda:

- Many coal-fired power plants were built in recent decades in rapidly growing economies. While the average coal plant in the US or Europe is forty years old, in developing countries it is fifteen. Their long remaining life span is a drag on decarbonization without CCS.
- For coal plants that are not closed, and for natural gas generation, CCS may cut emissions 90% or more while providing baseload power that supports reliable wind and solar integration.
- Many industrial needs cannot be met with renewables, and must use fossil fuels as inputs into production.
- Net zero by mid-century almost certainly will require *negative* emission technology to offset remaining emissions. Bioenergy with CCS (BECCS) uses biomass to absorb CO_2 as it grows, generates power or heat with the biomass, and captures and stores emissions, creating "a closed cycle with a slightly positive carbon balance."[25]

There may come a time when CCS is no longer relevant for coal-fired power plants and use with natural gas is unnecessary. However, for many industry applications, and as a negative emission technology when combined with biomass, it should be part of clean energy discussions. It is for the hard-to-decarbonize industries (cement or steel) that the strongest case can be made.

The basics of carbon capture and storage

Many technologies remove pollutants from waste streams. Air pollution control is built on them, from catalytic converters for cars, to scrubbers for removing sulfur dioxide from coal plants, to baghouses that control particulates.[26] Indeed, pollution laws typically rely on technology-forcing: define the best available technologies and make pollution sources install and operate them. From time to time, new technologies are required for more stringent controls.[27]

With this strategy, much of the world reduced industrial and transport pollution. In many ways, however, it had the effect of reinforcing carbon lock-in by minimizing the social costs of fossil fuels and making them tolerable. Whatever we may think of air pollution around the world, it would be much worse without these technologies. California deserves much of the credit for stimulating the development and use of these technologies.[28]

The rising concern with climate change has challenged this model. It is causing us to punctuate the policy equilibrium. Removing CO_2 from waste streams is more difficult and costly than for other pollution control. The sheer scale of emissions means that there are prodigious amounts of CO_2 to dispose of, safely and without leakage, for a long time. Add to this the residual effects of fossil fuels on public health and ecosystems, and we have the case for a clean energy transition.

The technology for removing CO_2 from fossil fuel combustion is, of course, carbon capture and storage.[29] There are other ways to reduce coal emissions, such as technologies for generating electricity from coal more efficiently, but they come nowhere close to what is needed to avoid the worst impacts of climate change by 2050. CCS refers to technologies that remove carbon dioxide from air waste streams. Efforts to capture and store CO_2 in the US go back to the 1970s oil crisis, when it was used for enhanced oil recovery.[30] CCS has three steps: capturing the CO_2 and converting it to a compressed gas or pressurized liquid for transport; moving it in pipelines, on ships, or by road; and storing it deep underground in geological formations: saline aquifers, depleted oil/gas fields, or coal seams. Capture occurs pre- or post-combustion or through an oxyfuel combustion process. Transport is costly, so it helps to locate facilities near disposal or capture sites. Storage poses technical challenges, given the need to prevent carbon leakage for indefinite periods of time.

The Carbon Capture and Storage Association, an industry group based on London, and obviously having a bias toward the technology, asserts that "At every point in the CCS chain, from production to storage, industry has at its disposal a number of process technologies that are well understood and have excellent safety records."[31] This may be true, up to a point, but the CCS picture is not quite so rosy. Cost is the main issue: CCS struggles to be competitive, especially given the rapid decline in wind and solar costs. Other issues are residual CO2 releases, the effects of other air pollutants, and worries over leakage from disposal sites.

When the IEA published its major assessment of CCS in 2016, it celebrated a milestone: the twentieth anniversary of the world's first dedicated CCS plant. For two decades, Norway's Sleipner plant had been separating CO2 at a natural gas site, processing it, and injecting it deep underground in a sandstone formation. It was a *dedicated* CCS plant because the injected CO2 was not used for enhanced oil or gas recovery, which three previous US plants had been built for. With enhanced oil recovery (EOR), CO2 has an economic purpose: it creates pressure for withdrawing oil and gas that otherwise is inaccessible. The economic value of the oil and gas helps offset CCS costs. EOR (and EGR, enhanced gas recovery) makes CCS economically feasible. The irony is obvious. It uses fossil fuel production to cross-subsidize carbon removal, but that makes it work.

CCS critics in the US claim the technology has not been proven commercially. The primary evidence, and a word that clouds debates on CCS, is *Kemper*. Located in Kemper County, Mississippi, and developed by the utility company Southern Company, this CCS plant was "so technologically complex its builder compared it to the moonshot of the 1960s."[32] It was a big part of the Obama climate plan, a planned job-creator, and would have been the world's largest CCS plant. Originally expected to cost $2.4 billion, it had mushroomed to $7.5 billion by 2017, when the Southern Company scuttled it "after years of blown-out budgets and missed deadlines."[33] Shoddy construction, poor planning, design flaws, and falling natural gas prices, which also affect nuclear and renewables, combined to make the Kemper plant economically unworkable.

For critics, Kemper proved that, even with subsidies, CCS was not feasible economically. Others disagreed, including well-known environmentalists. David Hawkins, of the US Natural Resources Defense Council, argued that Kemper's problems should not reflect broadly on CCS, and cites other projects that delivered on time and on budget.[34] Successful counterpoints are Boundary Dam in Canada and WA Parish in Texas (also known as Petra Nova). Built as a modernization and retrofit of an existing plant costing $1.5 billion (of which $800 million went to CCS), Boundary Dam captures up to 90% of CO2. Twice as large, the WA Parish Generating Station/ Petra Nova facility captured 90% of emissions that were sold to Gulf

Coast oil fields for EOR. These were the only operating coal-fired CCS plants in North America until Petra Nova closed in 2020, a victim of falling oil prices.[35] One critic argues that "both are really only demonstration units."[36] With 2018 costs triple those of coal without CCS, and many times that of wind and PV, even accounting for renewable storage costs, the CCS case for coal plants is unproven, at best.

Does CCS make sense?

Critics of CCS argue that it perpetuates carbon lock-in, is unproven at scale, and diverts scarce investment from such priorities as wind and solar, battery storage, hydrogen, infrastructure, EVs, and nuclear. CCS defenders say that it has not really been tested commercially at scale, is needed given remaining life spans of coal-fired power plants in many countries, and fills the transition needs of helping balance grids and serving industry heat and feedstock requirements in coming decades.

The need for CCS also depends on our climate ambitions. As the World Resources Institute noted a few years ago, based on a UN report, the world will not get to a 1.5-degree goal without removing and storing some eight gigatons of CO_2 with CCS each year. This has to be offset beyond natural removal gained through "improved management of forests, wetlands, grasslands, and agricultural lands."[37] Even meeting 2 degrees may require aggressive use of CCS, CCUS, and BECCS.

We can aspire to the energy system of our dreams or to what is realistic and affordable. Most decarbonization scenarios anticipate some reductions coming from CCS. In 2016, the IEA stated that the "transformation of the power sector will be *at least* USD 3.5 trillion more expensive" without it.[38] Without using removal technology on a large scale, the IEA and others are saying, the world will need far more wind and solar. As a practical matter, governments and investors are unlikely to tolerate the stranded assets involved in closing plants with long remaining life spans. For electricity and heavy industry, it is hard to take CCS out of the picture, despite its problems.

With aging coal plants and the capacity to expand renewables, North America, Europe, and other developed countries should be aiming to shut down coal-fired generating plants very soon. If CCS contributes, as most experts think it must, it will be as a supplemental technology in hard-to-decarbonize heavy industry sectors like steel and in cleaning up natural gas. Given the long remaining life spans of coal plants in developing countries, CCS may be a needed technology.

Public policy has a role in lowering costs, improving efficiency, and creating a transport and storage infrastructure. A start is pricing carbon with a tax or cap-and-trade policy. As long as we do not account for the social costs of fossil fuels, there are few economic reasons to invest in CCS. A sound carbon price changes that calculation. Other policies will be necessary, including those that target CCS directly: research and development, safety standards for coal and gas plants, labels on carbon content, and tax credits. As with any policy, "Long term commitment and stability in policy frameworks is critical."[39]

Negative emissions: bioenergy with CCS

Building a decarbonized global energy system is a formidable challenge. Yet we are at a point where even a largely decarbonized system later this century will fall short of our climate goals. Almost all the leading scenarios now incorporate carbon removal technologies, with two purposes. One is to offset residual emissions, such as from transport, buildings, or heavy industry, since the odds are that the carbon will not be squeezed entirely out of these hard-to-decarbonize sectors. Second, and related, is to compensate for a temporary *overshoot* of emissions and get the world to net zero.

As noted earlier, nature provides options for removing carbon from the atmosphere. But even assiduously pursuing those is not enough, and technology has a role. One option is direct air capture, or extracting already-released carbon. Given our focus on the energy system, direct air capture is not taken up here, except to note that it has potential. But costs and scale are constraints, making it a last although perhaps necessary option.[40] One technology for negative emissions that is directly linked to the energy system is BECCS: bio-energy with carbon capture and storage.

The strong, practical case for this technology, at least beyond the short-term, is negative emissions – getting not only near zero but also removing carbon. Many options exist for carbon removal: stopping deforestation, reforestation and afforestation, new methods of soil management and agriculture, direct air capture. BECCS combines the carbon sink (that is, negative emissions) aspect of growing biomass with the almost carbon-neutral features of advanced CCS to achieve net carbon reductions while also meeting energy needs. Nature is harnessed to technology.

BECCS requires an operation that uses fossil fuels for an economic activity. A power plant is a candidate, but combined heat and power plants, pulp and paper mills, lime kilns, ethanol, and waste-to-energy plants may also work. Access to biomass is essential. According to the Global CCS Institute, of the five BECCS plants operating in 2018, the largest is linked to an Archer Daniels Midland corn-based ethanol plant in Illinois. It captures and stores 1 million tons of CO_2 a year. The other four, in the US and Canada, together store half that amount. Three other plants that were under development at the time of writing were in Japan and the UK, both power plants, and in Norway, the last taking emissions from waste-to-energy and cement plants.[41]

Current BECCS thus accounts for 1.5 million tons of carbon capture and storage each year. To play a meaningful role in global decarbonization by later this century, the IPCC's 2014 report sets a target of 3.3 gigatons annually, *a 2,200-fold increase from current levels*. This imposes huge demands on land and water. Reaching this target, the Global CCS Institute admits, requires a land area twice the size of India and double the water used in agriculture. It increases fertilizer use by a factor of twenty, with profound effects on water quality. "The limiting factor of BECCS is not technology," the Institute notes, "it is the supply of biomass."[42]

Another problem is that BECCS is most likely to work if linked to bioethanol production. There are times when bioethanol makes sense, if made from certain waste crops, for example. But corn-based ethanol is full of problems given the land, water, and fertilizer used, all with damaging effects, and a low energy return on that invested (EROI). Estimates put the EROI for corn ethanol as low as one-to-one, which means that it takes as much energy to produce it as is gained. A 2018 study found wide variation in the EROI of biomass sources, with some falling below even a 1:1 ratio.[43]

Whatever the specifics, BECCS depends on CCS. It removes CO_2 from the atmosphere as well as from emission streams; this is a plus for bioenergy. Still, if *CCS does not meet the tests of cost-effectiveness, public acceptance, and safe transport and storage, then BECCS has nowhere to go.*

Yet the case for carbon removal, beyond natural tools like affor-estation, is compelling: "somewhere between 15–40% of the CO2 that humanity emits will remain in the atmosphere for up to a thousand years, with 10–25% of it persisting for tens of thousands of years."[44] Indeed, of the 116 scenarios the IPCC considered in its *Fifth Assessment Report* in 2014, 101 called for negative emissions later this century. One study, concerned that carbon removal's prominence in scenarios is "a dangerous distraction," also concludes that "we are not in a position to discard the negative option easily."[45]

Is CCS, and BECCS in particular, a distraction from the clean energy transition or, even more worrisome, could it be a delusion that fails to deliver? CCS probably has not proven itself as a commer-cially viable option on a large scale, except maybe when linked to enhanced oil and gas recovery, which eventually will be unnecessary in a decarbonized world. It may help ease a transition for the oil and gas industry. BECCS as a method for delivering negative emissions is even less proven commercially, and its link with environmentally damaging bioethanol is a problem. Writing in *Wired* in 2017, Abby Rabinowitz and Amanda Simson argued that the Paris Agreement and the IPCC are betting on "a technology that basically doesn't yet exist." They view CCS as "a conceptual tool, not an actual technology that anyone in the engineering field ... is championing." They ask: "Has the world come to rely on an imaginary technology to save it?"[46]

Hydrogen and Clean Energy

Renewable and nuclear energy are not the only alternatives to fossil fuels. Hydrogen is the most abundant chemical substance in the universe and the lightest element in the periodic table, but it has to be separated from other elements for practical application. For clean energy purposes, this occurs through a process of electrolysis, "which splits water into hydrogen and oxygen using electricity from multiple energy sources."[47] The issue is how that electricity is generated – by fossil fuels or renewables. Hydrogen produced with fossil fuels does not deliver a decarbonized world.

Hydrogen is not an energy source but, like electricity, an energy carrier.[48] Its advantage over electricity is that it is a chemical energy carrier, which means that it can store energy for long periods, is transportable, and may be burned at high temperatures to produce heat for industry. The key to its having a role in clean energy is how it is produced. As with electricity, a primary energy source is needed to produce it. Only about 5% of hydrogen is produced through renewables, known as *green hydrogen*. The rest uses fossil fuels, usually natural gas but sometimes coal. It is termed *grey* hydrogen or, if emissions are captured, *blue* hydrogen.[49] The opportunities lie in ramping that 5% of green hydrogen to or near to 100%, perhaps with some of the blue variety to support a transition. Hydrogen potentially meets multiple needs in a clean energy system:

- It may store excess sun and wind energy as a gas or liquid for longer periods than batteries, making it valuable for integrating variable renewables into electricity grids.
- It supplements EVs in low-carbon transport. Fuel cells mix hydrogen and oxygen to produce electricity that powers vehicles. Water is the only by-product. It may complement EVs and someday offer solutions for heavy-duty, long-distance transport, including ocean shipping and aviation, where batteries are unfeasible.
- Hydrogen fills a gap by producing the heat needed for cement, chemicals, iron/steel, and other heavy manufacturing. Green hydrogen may at some point be more economical than fossil fuels with CCS. It could work as feedstock (input to production) in replacing fossil fuels.
- It blends well with natural gas and meets gas needs with lower emissions. Experts say it is feasible to mix 10–20% of hydrogen with natural gas to create less carbon-intensive fuels and feedstocks. This helps with many hard-to-decarbonize sectors. Mixing with gas creates economies of scale that lower costs to support green hydrogen deployment.

Hydrogen complements our two beyond-renewables technologies – nuclear and CCS – by facilitating the integration of renewables and enabling time shifting of energy. Nuclear and CCS provide baseload generation, and hydrogen expands storage capacity. All three, CCS and hydrogen especially, can generate industry heat with far lower or no carbon emissions. Hydrogen fuel cells also expand electricity options for long-haul commercial road transport.

Chemical storage

A prime advantage of hydrogen is that it offers a means of chemical energy storage.[50] While renewables provide a *flow* of available energy, we know that flows rise and fall by time of day and season. Green hydrogen generated through electrolysis complements wind and solar with a *stock* of energy that can store electricity for long periods.

Integrating renewables like wind and solar into electricity grids requires huge amounts of storage. Batteries are part of the solution, as discussed in the previous chapter; hydrogen is another. Because of its low density, hydrogen requires space, which is less of a problem for stationary storage than for vehicles. Hydrogen's role is in using excess generation at peak times to power electrolysis that produces hydrogen for producing electricity later, when needed. While current technologies for battery storage are limited to about four hours, hydrogen can store energy for days or weeks. Of course, there are energy losses. And it must be stored somewhere – underground caverns are one option – then transported and converted back to electrical energy. Again, the issues are technology costs and economies of scale.

In the longer term, perhaps decades, hydrogen storage may occur with fuel cells, like those used in vehicles, but with larger storage capacities. The main issue now is the cost of developing technologies at scale for stationary fuel cells. In the near term, green hydrogen helps decarbonize electricity when mixed with natural gas. The IRENA finds that a 10–20% mixture of hydrogen with natural gas could be accommodated within existing infrastructure and uses; going above 20% involves more adjustment and higher costs.[51] Mixing hydrogen and natural gas lowers emissions. Near term, it expands hydrogen demand, helps economies of scale by keeping electrolyzers busy, and cuts the unit costs of producing hydrogen.[52] Blue hydrogen is a path to green hydrogen.

Gas mixing, an incremental step, thus paves the way to more disruptive change. As the IRENA puts it, mixing is a "low-value, low-investment stepping stone to support the early-stage scaling-up of production" and traversing the valley of death that plagues new technologies.[53]

Currently, pumped hydro and compressed air storage cost less than green hydrogen, but this can change.[54] Hydrogen avoids the topographical and geologic constraints of these older technologies and works longer term, with less chemical degradation than batteries. To illustrate its flexibility, hydrogen could transport electricity from offshore wind platforms via pipelines to population centers. It generates as well as stores clean electricity. An example is Green Hydrogen Hub Denmark, which aims to "secure renewable energy regardless of the weather." It combines one of the world's largest electrolysis plants, 200,000 MW of underground hydrogen storage, and compressed air storage that will deliver electricity and meet multiple industry needs by 2025, displacing 600,000 tons of CO2 emissions a year.[55]

Low-carbon mobility

In the US in 2019, there were 7,500 fuel cell vehicles in use (most of them in California, with its incentives and charging stations) out of 275 million vehicles. Globally in 2019, about 11,000 fuel cell vehicles and 7 to 8 million EVs were in use, out of a global total of about 1.4 billion vehicles. Do the math, and it is clear that hydrogen is not taking over the roads.

Hydrogen is applied to mobility with the use of a fuel cell: "an electrochemical reactor that converts the chemical energy of a fuel and an oxidant directly to electricity."[56] Its uses depend on the circumstances. In densely populated Japan, for example, charging stations may be concentrated by location. The country also depends heavily on costly imported oil. Japan is aiming to "create the first 'hydrogen society,'"[57] with a goal of producing 200,000 fuel cell vehicles by the mid-2020s. Automakers recognize the potential. Hyundai aims to produce 700,000 fuel cell vehicles annually by 2030.

Fuel cell mobility shares features with EVs. An obvious one is clean energy – although in both cases this depends on how the electricity powering the cars or producing the hydrogen is generated. Both are efficient for converting energy into movement, with low maintenance costs. But they also share disadvantages, including price, charging infrastructure, and consumer acceptance. Since these disadvantages are currently more pronounced for fuel cells than for EVs, in most countries the latter will dominate, at least for a time. Still, hydrogen could play a larger role, especially for trucks and buses.

Buses illustrate hydrogen's mobility strengths. Urban buses are better suited to hydrogen than to electrification, one expert writes, because "they need heavy battery packs for electrification and tend to travel to and from central bases where it could be economically viable to build a hydrogen refueling station."[58] This holds also for trucking: large trucks that go long distances could refuel at stations on set routes. This is similarly the case for small commercial vehicles such as delivery vans or garbage trucks. When the technologies and economies of scale are right, hydrogen may work for trains, local water transport (such as ferries), and to complement aviation biofuels.

Although commercial vehicles are the near-term target, governments have also set goals for passenger vehicles. The 2019 *Hydrogen Roadmap Europe* forecasts that, by 2030, the world will have fuel cells in 3.7 million passenger vehicles, 500,000 light commercial vehicles, and 45,000 trucks and buses, along with 3,700 large refueling stations and 570 fuel cell trains. It expects hydrogen to replace one-third of natural gas use by 2040 and industry to greatly expand ultra-low-carbon hydrogen production by 2030. As noted earlier, Japan is committed to hydrogen mobility, aiming for 800,000 fuel cell passenger vehicles by 2030. Toyota introduced the first commercialized fuel cell car (Mirai) in 2014, which was followed by Honda's Clarity.[59]

Hydrogen vehicles currently cost 40% more than EVs and 90% more than internal combustion cars.[60] At this point, purchase costs are higher because fuel cell systems and components are expensive, an issue that economies of scale eventually should solve. Operational costs are also higher because hydrogen is expensive. Projecting less costly technology and "dramatic improvements" in the economies of scale, Deloitte thinks ownership costs for fuel cell vehicles will fall by half by 2030.[61]

Industrial heating and feedstock

Among the "really hard stuff" discussed in Chapter 3 were industrial energy demands, in terms of both heat source and feedstock for production. Feedstock use currently accounts for the bulk of hydrogen demand around the world, almost all of the grey variety. Ammonia production, mostly fertilizers, makes up 60% of industrial hydrogen, and petroleum refining 23%. These will continue and help drive economies of scale.[62]

Hydrogen has two industry applications. One is as a feedstock for producing chemicals like ammonia. The other is as a heat source for carbon-intensive processes. Hydrogen "has the technical potential to channel large amounts of renewable electricity to sectors for which decarbonization is otherwise difficult."[63] Wind, solar, and other renewables are not suited to producing industry heat, but hydrogen produced with electrolysis may at some point generate industry-level heat as effectively as fossil fuels do now.

Many factors are driving the interest in hydrogen. On the positive side are the falling costs of wind and solar; progress in electrolysis technology, where costs fell 60% from 2010 to 2019; more flexible electricity grids; growing economies of scale; potential use in difficult sectors (aviation, heavy industry, ocean shipping); and governments' net-zero commitments. On the other side, production costs are still high, lots of infrastructure is needed, energy is lost all along the value chain, and green hydrogen demand is insufficient.[64] Carbon pricing would help. Financial incentives for accelerating EV adoption also apply to fuel cells, as do mandates aimed at industry.

In summary, the opportunities start with the use of green hydrogen in chemicals, petroleum refining, iron and steel, and other heavy industry. Next is freight transport – trucking, rail, shipping, aviation – where fuel cells may have an advantage. Also on the menu are heat generation for industry, high-use transport (taxis, light-duty delivery trucks), and green hydrogen to enable flexible grid storage.

Beyond Renewables?

Each of these technologies – nuclear, CCS, and hydrogen – have a role in the transition. Just *what* role and *when* are the questions. Existing nuclear delivers reliable, carbon-free baseload power, at least in operation. The economics are a problem. A larger one is risk perceptions. In the meantime, there is a case to be made for continuing to invest in small, modular, passive safety reactors. Table 7.2 presents a summary of the contributions, barriers, and factors facilitating scale-up for the technologies discussed in this chapter.

The IEA calls nuclear and hydro "the backbone of low-carbon electricity generation" in the world, together accounting (in 2019) for three-fourths of it.[65] What is problematic is the prospect of a "nuclear fade," in which plants close and renewables cannot take up the slack. The IEA urges that prudent steps be taken to preserve and expand nuclear's role: keep existing plants for as long as they can be

Table 7.2: Summary of the Three Complementary Technologies

Issue	Nuclear Energy	CCS/CCUS/BECCS	Hydrogen
What it could contribute	Reliable baseload power Carbon-free source at scale Flexible, cost-effective source (new models)	Clean up coal plants' remaining life Enable fossil fuels in industry for the transition Cleaner gas during the transition Negative carbon potential	Long-term electricity storage; ramp-up potential Fuel cells: use in commercial heavy transport, buses Ocean shipping and aviation Feedstock and heat for industry
Issues with & barriers to scale-up	Costs of current technology (capital) Safety, waste, security concerns Public perceptions of the technology	Not proven commercially at scale Depends on EOR/ EGR Links to ethanol plants; land use for BECCS scale-up Still emits carbon/ pollution	High costs for green version Lack of dedicated infrastructure Energy lost in value chain Challenges in storing & delivery Low mass density
Factors facilitating scale-up	New generation of reactor technologies Widespread recognition in clean energy scenarios New passive safety designs	Remaining life spans of coal and gas plants Political support for fossil fuels Need for some baseload power, industry transition	Falling wind/solar costs Technology ready for scale-up Demand in really tough sectors High energy density
What needs to happen	Small, modular reactors develop at scale Investors and public need to be convinced Licenses for existing plants extended as long as safe Potential uses in industry explored (e.g. hydrogen)	Costs have to come down Regulatory framework needed for safety & reliability Utilization expands to help the economics Land and other issues of BECCS are overcome	Wind/solar costs fall even more Fuel cell technology improves Infrastructure built (hydrogen hubs) Demand forces a scaling up of technology Electrolysis costs decline

safely operated; value nuclear dispatchability in markets; recognize the environmental and security benefits through carbon pricing; use innovative financing; maintain human capital (knowledge and experience) in nuclear; and invest in new reactor designs that are safer and have low capital costs.

A strong case for CCS is the existence of coal-fired plants with many years of useful life remaining. Even if no new plants are built, the existing ones would pump out CO_2 for decades. Coal plants are well on their way to becoming history in developed countries, on environmental and cost grounds. But rapid-growth economies need to power industry and lifestyles, and coal is there. The use of coal and, and more importantly, natural gas in industry also makes a strong case for CCS. Eventually, fossil fuels will be phased out. BECCS draws attention, particularly for its negative carbon potential. Its association with bioethanol is not a mark in its favor, given its low EROI and harmful environmental impacts.

Green hydrogen does not carry the same baggage as the other two technologies, but it has a long way to go. With only 5% of electrolysis driven by renewables, a major scale-up is needed. Green hydrogen is a versatile technology capable of serving multiple purposes: powering mobility, storing energy, generating heat. There are few easy options in the transition, but green hydrogen is close to being one. CCS has a shorter-term role, although carbon removal will be with us long term. Nuclear remains in question, given the uncertainty about new technologies. The IEA rates CCUS and nuclear as *not on track*, while green hydrogen earned a more positive *more effort needed*.[66]

The transition cannot be leisurely, shaped by market vagaries and unguided accumulations of choices. Only if governments combine to drive it with well-designed, coherent policies will the technologies advance, the economies of scale emerge, and the many elements of a complex system be made into a successful whole. The tools available for government are the subject of Chapter 8.

Guide to Further Reading

Robin Cowan, "Nuclear Power Reactors: A Study in Technological Lock-In," *The Journal of Economic History* 50 (1990), 541–67.
Howard J. Herzog, *Carbon Capture*, MIT Press, 2018.

International Energy Agency, *Nuclear Energy in a Clean Power System*, 2019.

International Renewable Energy Agency, *Green Hydrogen: A Guide to Policy Making*, 2020.

8

Accelerating the Energy Transition

Norway has by far the highest percentage of electric vehicle sales of any country. It is worth asking why. Norway reveals a great deal about how public policy can accelerate the pace of an energy transition. Its path to an electrified passenger fleet and goal of banning sales of new internal combustion engine vehicles by 2025 demonstrates how policies may be designed to achieve clean energy goals.

The roots of Norway's EV policies go back to the 1990s, when evidence of the threat of climate change was beginning to draw serious attention. Norway was an ideal breeding ground for innovative EV policy. Its lacks its own ICE manufacturers, so there was no domestic industry to worry about. Nearly all of its power (96%) comes from hydro, providing clean electricity for charging. Norway has high vehicle license fees and taxes, and these have been used to offer financial incentives. It is an affluent country, so there were ample consumers for creating early niche markets.

Policies for promoting EVs began in 1990 with a temporary exemption from registration for experimental EVs.[1] This became permanent in 1996. Exemption from the value-added tax was introduced in 2001, and reduced annual license fees in the late 1990s and early 2000s. These incentives saved consumers thousands of euros. The government began providing direct subsidies with free access to toll roads and reduced ferry fees to promote EVs outside of urban areas. Consumers were also enticed with free and accessible parking and access to bus lanes. The government supported expansions in charging stations, including fast chargers, in 2009 and 2011. By

2015, EVs were 20% of new passenger car sales. By 2020, this had risen to nearly 75%, followed by Iceland at 45% and Sweden at 32%. The global average was only 4.2%, but growing.[2]

Norway's strategy, which emerged incrementally over decades, illustrates many of the policy tool categories discussed in this chapter. It relied heavily on *financial incentives* with tax and fee exemptions and subsidies through toll waivers, and on *mandates* by granting access to bus lanes and, due in 2025, a ban on new gas and diesel sales. The government also used another policy category tool by *creating a market* for EVs that, combined with other actions, such as EU mandates for low-emission vehicles, encouraged carmakers to expand their offerings.

Three big lessons emerge from Norway's experience. One is that policies matter. For EVs, "European markets with substantial incentives have higher market shares than those with less or no incentives."[3] Second, stable policies are helpful. As Eric Figenbaum writes: a "transition policy should be seen as lasting five, 10 or up to 20 years, which requires leadership, persistence and the ability to deal with unexpected events."[4] Third, policies can combine to have positive effects. Policies in Norway and California stimulated advances in battery technology, lowered costs, and created economies of scale in a virtuous circle of energy innovation.

Using public policy to accelerate the energy transition is the subject of this chapter. Public policies are actions governments take to achieve collective goals. At times, this involves changing behavior directly with regulation. At other times, it is about changing incentives, as with tax credits. The chapter begins with an overview of policy and the forms it may take. It builds on an earlier categorization of the options as *regulating, taxing, trading,* or *informing.*[5] I modify these to view them as *mandates, financial incentives and support, market creation or stimulation,* and *information disclosure.*

Public Policy and Clean Energy

Energy policy focuses on the capacity to do work beyond using human and animal muscle. Had there been an energy policy field in the late 1700s and early 1800s, it would have studied how water and wind generated energy for mills and ships, as well as coal's role in early industrialization. In the late 1800s, policy analysts would have been intrigued by the new technology of electricity and its economic and social consequences. In the first half of the twentieth century, the focus would have been on the significance of fossil fuels in promoting national economic and military power, and on oil in powering mobility. In the second half of the century, the policy field concentrated on the growth of nuclear power after World War II and the rise of natural gas, once the pipeline technology for transporting it was developed. By the end of the twenty-first century, the need to deal with climate change was central to energy policy, although issues of supply, technology, and economics in the existing fossil-fuel-based system still mattered. Now, a clean energy transition is the theme.

A major focus of the study of public policy is understanding the mechanisms government uses to change behavior to achieve a collective goal.[6] There are many such goals: cleaner air or water, low infant mortality, public safety, and so on. Among energy policy goals are dependable gasoline supplies, reliable electricity access, cleaner transport fuels, low-cost energy, and efficient mobility. The goal we are focusing on here is a largely decarbonized energy system and net-zero carbon by mid-century.

In decarbonizing, policymakers aim to change behavior from a business-as-usual path to one promoting the clean energy pillars of efficiency, renewables, electrification, and complementary technologies. The chapters so far have suggested several mechanisms for changing behavior:

- *Efficiency*: building energy codes, efficiency resource standards, appliance standards
- *Renewables*: utility obligations, feed-in tariffs, technology subsidies, tax credits
- *Electrification*: EV tax incentives, support for grid flexibility, smart grids, ICE bans
- *Complementary technologies*: research, demonstrations, tax incentives, carbon pricing

These are but a sample of the tools available. All aim to accelerate the transition, because the costs of relying only on market forces or loosely connected policy interventions are too high. At least that is what many governments think. This energy transition is more intentional than previous ones, and will be driven by a goal of minimizing energy's social costs, especially climate change.

This is not to say that other criteria are not important. People depend on energy and are sensitive to its costs. Equity is an issue, whether in relation to electricity access in low-income countries or energy burdens in rich nations. If policies are too costly or perceived as unfair, they will falter. We expect policies for a transition to get the intended results. Climate change especially offers limited options for do-overs. If global energy does not change from the business-as-usual path by mid-century, the harm will be irreversible.

Energy costs are a constraint on policy designs. Spain's ambitious clean energy transition was almost derailed by moving too quickly. Incentives for investing and expanding wind and solar were so generous that government budgets and costs got out of hand. California's transition has had to pay attention to integrating wind and solar into grids. Infrastructure issues are a constraint as well: Kenya's plan to harness its geothermal resources is limited by the existing transmission capacity. Political interests, a cause of carbon lock-in, pose challenges in places like Australia and the US.

For the mechanisms used to change behavior, I use the term *policy tools*. A tax credit is a policy tool, as is an energy efficiency obligation, a subsidy to advance storage or carbon capture, or a mandate that utilities must derive a percentage of generation from wind, solar, or other renewables. Linking multiple tools in coherent, coordinated ways leads to *policy strategies*. The term *strategy* is a flexible, scalable concept. Each scenario outlined in Chapter 3 implies a strategy, as do the combinations of tools used to advance efficiency, expand electricity access, or accelerate EV adoption.

Putting Clean Energy Policies in Context

Many issues matter in analyzing clean energy policy tools. Two worth discussing at the outset are the stage of development of a technology or practice, and the role government should play.

Stage of development

A consideration in policy design is the stage of development of a technology or practice. Early on, the priority is scaling them up so they may become competitive.[7] Government plays a role in sponsoring prototypes or demonstrations to prove feasibility and create market demand. Policy can adapt by creating a degree of regulatory certainty about how a technology will be managed. With advanced nuclear, for example, it helps to streamline reactor licensing. CCS will progress only with some regulatory certainty on licensing and approval processes. To promote investment in technologies, government can provide assurances about how they will be regulated, and support scale-up through demonstration projects or more directly by stimulating market demand.

At a more mature stage, once technologies are shown to be feasible, competitive, and able to attract investment, subsidies are less necessary. The task becomes one of creating policies for rapid scale-up. Carbon pricing supports the rapid diffusion of renewables by increasing the costs of carbon fuels and moving utilities, industries, consumers, and investors toward cleaner sources. Delivering support via subsidies or incentives matters less at this stage than reorienting economic calculations broadly toward less carbon-intensive energy. Relative costs make a difference. Once EVs reach parity with ICEs, tax credits and other incentives are less necessary. EV's virtues – low maintenance and fuel costs, expanded range, diverse recharging options – make them competitive on their own.

This chapter focuses on elements of clean energy that are now or could be competitive, and that can be scaled up to play a role by mid-century, though not all are a sure bet. The chapter does not address research and development, although that is critical not only for emerging technologies but for aspects of existing ones. My focus is on policy tools that get us to net-zero carbon by mid-century using mostly existing technologies.

The role of government

Much of what governments do to advance collective goals involves spending money. In energy, key areas for public spending are research and development, infrastructure funding, and market scale-up. These are *public goods* where the private sector has little incentive to take investment risks. Government funds research where the prospects of financial return are small or nonexistent. Funding for infrastructure is crucial because energy systems depend on the shared ownership of critical physical systems: transmission networks, pipelines, modern grids, and so on. These involve high capital costs. President Joe Biden's plan for 2 trillion dollars in infrastructure investment that was announced in 2021 included clean energy options, as do other countries' energy plans.[8]

Government also plays a role in scaling up promising technologies by helping them get through what is known in innovation studies as the *valley of death*, where technologies never move beyond the research or demonstration stage to commercial success. Public investment can support promising technologies or practices by creating market demand to facilitate scale-up or by sponsoring demonstrations to provide evidence of economic feasibility.[9]

Yet public investment alone is insufficient. Government money is essential, but the investment needs are so large that most funding must come from the private sector. The IRENA calls for total clean energy investments of some $3.5 trillion each year, nearly 4% of current global GDP.[10] It also makes sense to draw on the private sector to decentralize investment decisions. Having private firms help determine investments brings multiple experts, motivations, and perspectives into the equation. Decarbonizing requires government participation and guidance, but leveraging private actors and using market forces leads to better outcomes. My focus here is less on funding than on how governments may use various policy tools to deliver the investments, behavioral change, and other actions needed to decarbonize global energy.

Policy Tools for Clean Energy

We will examine clean energy policy tools in four categories. The first is *mandates*. The most direct, if not always most efficient, effective, or equitable, way to change behavior is to issue directives and back them up with sanctions. Most regulation consists of mandates in various forms. These come as prescribed technologies or as directives to achieve a level of performance, such as a minimum of air pollution or maximum of efficiency for generating electricity from coal plants.[11]

Second is changing behavior with *financial incentives and support*: investments, taxation, and subsidies.[12] This encompasses many policy tools, from direct public investment in research and innovation, to carbon taxes, to tax credits for wind energy developers. Their goal is to change economic calculations so they will lead to socially preferable choices. This includes *incentives* as well as *disincentives*. Examples of the latter are financial tools that penalize behavior that does not promote policy goals, such as carbon taxes.

A third category is to *create or stimulate markets*. One cause of our climate and pollution problems is market failure. Carbon dioxide, nitrogen dioxide, and particulate matter – not to mention water quality and ecological impacts – are not accounted for in markets. All the tools discussed here aim to correct market failures. This category does it explicitly. Government creates relationships or stimulates market demands that change behavior. Emission trading systems in the EU, California, China, and elsewhere are such government-created markets.[13] Governments also stimulate markets with public purchasing and certification programs.

Fourth is *information disclosure*, either by government itself or disclosure that is required of others. The approach here is not to issue a mandate to change behavior or realign choices with financial tools, but to correct a specific failure, that of information asymmetry. Investors, firms, consumers, voters, and others often lack the information required for making the desired choices, so government steps in to provide it. This alters behavior by appealing to consumers' economic self-interest (think of the money I will save with this refrigerator or by leasing space in a green building) or their values (this EV protects the climate and cuts pollution, so I should buy it for the good of the planet).

Policy experts propose other ways of differentiating policy tools. Some distinguish *market-push* from *demand-pull* policies.[14] This has its uses, such as distinguishing policies for increasing renewables through public investment and subsidies from those that expand demand, such as tax credits for EVs. But it often breaks down in practice; subsidies, for example, can offset the costs of investing in renewables (supplies) as well as the costs of buying them (demand). Another approach is to lump together tools like carbon taxes and trading, often with tax credits and subsidies, as *economic* or *market* incentives, different from regulation. Yet all these tools involve regulation that constrains behavior but does so in different ways. Simply describing a tool as *regulatory* does not take us very far.

The following sections examine policy tools for promoting a transition in the four categories outlined above. The list of clean energy tools is long, so I select here those that are used widely and build upon the clean energy pillars of efficiency, renewables, and electrification (chapters 4, 5, and 6).

Mandates: tell them what to do

We all accept that, to some degree, government has to constrain individual behavior for the sake of the greater good. Without a system of rules, markets would be dangerous and chaotic. Basic financial transactions would be unreliable if not impossible; rules for owning property and exchanging goods would be confusing and contested. Indeed, the foundation of social prosperity and social well-being lies in having systems of rules in place for regulating behavior.

The lines between individual choice and government regulation are constantly being disputed and renegotiated. Although few challenge rules for property rights or contracts, appliance standards or bans on inefficient lighting are contested.[15] Yet mandates backed by sanctions are a standard clean energy policy tool. Some prescribe standards of performance; others set technology or management standards. Three widely used types of efficiency mandates are building codes, appliance standards, and energy efficiency resource standards. All were introduced in Chapter 4 and are anchors for improving energy productivity and reducing energy intensity in years to come. As mandates, they set performance standards to meet or technologies and practices to adopt.

Efficiency should be easy, given the returns on most investments. Yet, as we saw in Chapter 4, 20-dollar bills are left lying on streets. The role of policy is to make people aware of cost-saving options and assert a public interest in reducing energy intensity.[16] Buildings account for one-third of global energy consumption and are used for decades, so they are a ripe target. Most experts think advanced designs can make new buildings at least 30% more efficient, an investment that will pay off not only financially but also in terms of less energy use and lower social costs. Codes "specify minimum energy efficiency standards for the residential and commercial building sectors" and cover the building envelope, lighting, water heating, and heating/ventilation/cooling systems.[17]

Two widely used tools for promoting sources like wind and solar are renewable portfolio standards (energy utility obligations in Europe) and feed-in tariffs.[18] Both are mandates but apply to different targets. Renewable portfolio standards (RPS) and energy utility obligations are sometimes referred to as *quotas* that require utilities to obtain a minimum percentage or amount of energy annually from renewable sources by a specified date. Quotas are a mainstay of renewables policy in the US. Iowa adopted the first RPS in the 1980s, and now more than one-half of states have binding RPS targets.[19] As of late there has been competition to adopt the strongest RPS; Hawaii aims for 100% renewable electricity by 2045; California's goal is 100% zero-carbon electricity by 2045. Feed-in tariffs (FITs) are used widely, especially in Europe, as a tool to expand wind and PV. They guarantee a minimum price for electricity from qualified sources, usually for specified time periods.[20] Germany has relied heavily on FITs as a policy tool.

California's zero-emission vehicle (ZEV) program, copied by ten other states, illustrates another mandate.[21] It requires that a percentage of vehicles sold be ZEVs, which include plug-in hybrid, battery electric, and hydrogen fuel cell units. It is based on a credit system and requires that about 8% of sales be ZEVs in 2025. Though a mandate, it also has elements of the other categories, such as market creation and stimulation. It incorporates trading by allowing carmakers to carry over excess credits from year to year and to buy and sell them. It starts as a mandate but creates a credit market, making it less costly. ZEVs stimulate markets by forcing automakers to produce a minimum of non-emitting vehicles, mandating economies of scale.

Well-designed mandates work. Although building and appliance efficiency standards increase construction and product costs, they deliver savings with short paybacks. Energy quotas also work. Renewable portfolio standards in US states increase the amount of wind and solar generation, but not necessarily the proportion of electricity from them, due to rising energy demand.[22]

Does an RPS increase electricity prices? Looking at the overall costs of integrating renewables into electricity grids, including transmission and intermittency, one study found that "electricity prices increase substantially after RPS adoption" by a few cents per kilowatt-hour.[23] If the goal is cutting carbon, a tax or cap-and-trade is a more efficient way to go. On the other hand, an RPS is effective in expanding wind and solar, if not necessarily in increasing their overall share. Even with higher prices, the social costs of fossil fuels make an RPS worthwhile. And political feasibility matters; if an RPS is more likely to be adopted than a carbon tax, clean energy advocates take what they can get.

Financial tools: change their economic calculations

Mandates aim to directly change behavior. Some tools take another approach: change the economic calculations of consumers and investors in order to realign them better with the social goal of cleaner energy. Make desired behavior *less costly* while making unwanted behavior *more costly*.

A range of tools aim to change behavior with financial incentives. One heavily used option is tax policy. The primary purpose of taxation is to raise revenue to fund government operations. A secondary goal is to structure tax systems around other policy goals. Fossil fuels have always benefited from tax policies favoring incumbents – a form of carbon lock-in. In the US, the oil and gas industries enjoy the benefits of a depreciation allowance offsetting tax liability when they extract non-renewable resources, a perverse clean energy effect. Wind benefits from a Production Tax Credit (PTC) and solar from an Investment Tax Credit (ITC).

Tax incentives are a leading tool in encouraging EVs. They appear to work. The International Council for Clean Transportation (ICCT) found that incentives are "a statistically significant driver for increased sales in multiple studies."[24] When Denmark removed registration tax exemptions in 2017, EVs fell to 0.4% of sales; when Finland expanded incentives in the 2010s, EVs grew to almost 6% of sales. The ICCT recommends four design principles for electric vehicle incentives: offer them up-front at purchase and make them visible; commit to them for many years, showing their durability; extend them broadly, including to corporate and government fleets; and be clear and transparent.[25]

A financial tool drawing wide attention is a carbon tax. This is one of two approaches to pricing carbon and changing behavior.[26] The second is an emissions trading system (ETS), such as cap-and-trade, discussed later. Both put a price on carbon but do it differently. Financial incentives like a carbon tax send a signal to investors, businesses, and others that fossil fuels will cost more over time (usually due to increases in the carbon tax rate), making zero-carbon sources more attractive. As the World Bank puts it, carbon prices are a way for public policy "to incentivize low-carbon action and avoid locking in more fossil fuel intensive investments."[27]

In 2021, the World Bank's annual *State and Trends of Carbon Pricing* identified over sixty national and subnational carbon pricing systems covering 21.5% of global emissions.[28] Over thirty were taxes; the rest were emissions trading systems. Covering only one-fifth of the world's emissions clearly is not a basis for an effective strategy. A bigger problem is that less than 4% of the prices were at a level needed to meet the Paris 2-degree goal, which the World Bank calculates should be $40–80/ton of CO2e in 2020 and $50–100 by 2030. Only 3.8% of prices are in this range, with an average of about 2 dollars.

Sweden adopted one of the first national carbon taxes in 1991.[29] It is the most stringent in the world; in 2020 a ton of carbon dioxide was taxed at a rate of 110 euros ($123 US). Carbon tax revenues enabled the Swedish government to cut marginal income taxes. Between 1990 and 2017, Sweden's emissions fell 26%, while its GDP grew 78%. In addition, it cut emissions by turning to nuclear, biomass, and hydro and by expanding district heating fueled by municipal solid waste and wood residues. Sweden rated at the top of the 2020 *Climate Change Performance Index*.

Other countries with rates high enough to meet the Paris goals are Switzerland, Finland, Norway, and France.[30] Others come close, including Canada, where the federal government adopted a national backstop carbon price that reaches $35 US in 2022; provinces must adopt their own pricing programs.[31] The Canadian province of British Columbia adopted a carbon tax in 2008 that was reauthorized in 2019. It rose to about $30 US and has gone up since then. British Columbia cut emissions, reduced income taxes, and helped low-income groups, all without economic harm.[32] Carbon pricing has also been adopted by private firms; about half of the largest 500 globally have internal carbon taxes to motivate emissions cuts and allocate investment capital.

Table 8.1 lists carbon pricing initiatives by country or jurisdiction as of 2020. It includes both pricing options: taxes and emissions trading. The second is discussed in the next section as market creation. The factors influencing the effectiveness of carbon pricing are price per ton of equivalent (tCO2e), given in US dollars, and the percentage of emissions covered. Having low rates and limited coverage (Spain, Argentina) is less effective than having high rates and broad coverage (Sweden, British Columbia). Some areas (Japan, Spain) have high coverage with low prices. China has expanded carbon pricing with a national program now being implemented.

Table 8.1: Carbon Pricing Programs in 2021 (Country or Subnational Jurisdiction)

Jurisdiction	Type	Price (USD/tCO2e)	Percentage of Emissions Covered	Revenue Generated (USD)
Argentina	Tax	5.50	20%	Under 1 m.
Canada	Tax	31.80	22%	3.4 b.
Chile	Tax	5.00	39%	165 m.
EU	ETS	49.80	39%	23 b.
France	Tax	52.40	35%	9.6 b.
Japan	Tax	2.60	75%	2.4 b.
Mexico	Tax	0.40–3.20	23%	230 m.
Singapore	Tax	3.70	80%	144 m.
South Africa	Tax	9.20	80%	43 m.
South Korea	Tax	15.90	74%	219 m.
Spain	Tax	17.60	3%	129 m.
Sweden	Tax	137.2	40%	2.3 b.
United Kingdom	Tax	24.80	23%	948 m.
British Columbia	Tax	35.80	78%	99 m.
California	ETS	17.90	80%	1.7 b.
Quebec	ETS	17.50	78%	549 m.
RGGI	ETS	8.70	23%	416 m.

Note: Mexico's range varies by source. RGGI is the US Regional Greenhouse Gas Initiative of eleven states for trading power emissions. The last four listed are subnational; the others are national. China's program is just starting and is not listed.
Source: World Bank, *State and Trends of Carbon Pricing*, 2021.

Market creation and stimulation

If private markets fail to deliver a social good, then why not *create* a market? Like financial tools, market creation is designed to realign economic calculations. Emission trading systems, a leading form of which is cap-and-trade, are an example. As long as utilities, firms, and consumers do not have to account for the effects of their emissions, for their social costs, they lack incentives to do anything about it. A carbon tax is one way of holding them accountable.

Another way is to put a cap on emissions and make emission sources hold a permit for each ton of CO_2 released.[33] If the cap is working, emission sources or energy users pay extra for carbon-based fuels. In general, this moves them away from fossil fuels, now that the societal harm of carbon-based sources is reflected in prices, and toward non-carbon sources like wind, solar, water, and nuclear. By increasing fossil fuel costs, trading makes non-carbon options relatively less costly and financially more attractive.

California launched its cap-and-trade program in 2013 to cover generating plants and large industry, and has expanded it since.[34] Sources of emissions need a permit for each ton they emit. They may buy or sell permits or bank them for future use. They may offset up to 5% of emissions with US-based, verified offset projects in such areas as forest practices and methane capture. Permits are sold at quarterly auctions; the price in November 2021 was nearly $30.00 per ton. By 2020, the state reported, the program had funded cumulatively over $5 billion in climate-related projects, over half of which went to low-income, historically underserved communities.[35]

Half the initiatives surveyed by the World Bank use market creation, usually a form of cap-and-trade. Government sets an emissions limit (the cap); allocates permits (a license to emit a ton) by auction or other means; allows trading and banking for future use; and requires sources of emissions to hold a permit for each ton emitted. Like California, they usually allow limited offsets – verified projects outside of the trading area to cut emissions or create carbon sinks.

Market *creation* is one set of tools; market *stimulation* is another. Clean energy is about more than inventing new technologies or practices. They must be *scaled up*. Having PV make up 5% of electricity or getting EVs to 10% of sales does not get the job done. Scale may be the most essential word in the energy transition. Larger scale usually leads to lower costs. The term *economies of scale* captures this.

A powerful tool in creating markets is procurement. The US government, for example, has some 650,000 vehicles, including those for postal delivery. State and local governments account for another 4 million. In January 2021, President Biden issued an executive order directing federal agencies to devise plans for converting federal, state, and local fleets to zero-emission vehicles. Given that less than 1% of government fleets were electric, this was "a boon to the fledgling electric vehicle industry."[36] Getting gas and diesel cars off the roads not only helps the climate, it expands the EV market and creates economies of scale.

Information disclosure: guide their choices

A classic market failure is *information asymmetry*, where people lack information for making rational choices. In economics, being *rational* means acting in ways that are consistent with self-interest. What could be more rational than not paying for unneeded energy? In addition, people often think beyond self-interest and use information to promote their values. They pay more for green products because they want to be socially responsible.

Whatever the motivation, governments can advance clean energy by giving information directly to consumers and investors or by requiring disclosure. Take appliance efficiency labeling. Japan's approach combines information with mandates. Its Top Runner program recognizes the most energy-efficient products in twenty-three categories and encourages manufacturers to earn this designation.[37] The EU's energy labeling and ecodesign program links minimum performance standards (a mandate) with a grading system (information) to encourage product efficiency. Labels "provide a clear and simple indication of the energy efficiency and other key features of products at the point of purchase."[38] Information tools tell us about the carbon content of products, reveal pollution liabilities in financial statements, and estimate lifetime vehicle emissions. Table 8.2 lists the leading policy tools by the categories defined here.

Lessons from California, Denmark, Spain, and Kenya

If anything is clear from the policy literature, it is that no one tool on its own can do the job. Countries making clean energy progress combine policy tools into coherent, coordinated strategies that are linked to long-term goals, often driven by numerical targets. At the same time, policymakers work with what they have. Resources, history, political preferences, and technology affect what is doable. Four settings for applying tools and designing strategies illustrate how this works.

California illustrates the use of multiple tools in combination. If California were a country, it would rank highly in the *Climate Change Performance Index*. Although its cap-and-trade system draws attention, there is a larger strategy that is driven by goals: reducing emissions from 1990 levels by 40% by 2030 and 80% by

Table 8.2: Summary of Clean Energy Policy Tools

Mandates
  Efficiency: building codes, appliance MEPS, vehicle efficiency standards, EERS
  Renewables: RPS/quotas, FITs, net metering
  Electrification: ZEV mandates, ICE bans, natural gas bans

Financial Incentives and Support
  Efficiency: carbon or gas tax, tax incentives, access to financing
  Renewables: carbon tax, tax incentives, subsidies
  Electrification: tax incentives for EVs and heat pumps, infrastructure investment

Market Creation and Stimulation
  Efficiency: cap-and-trade, tradeable certificates, government procurement
  Renewables: Cap-and-trade, renewable energy credits (RECs), government procurement
  Electrification: deregulation and restructuring of markets

Information Disclosure
  Efficiency: appliance labels, certification, audits, smart metering, benchmarking
  Others: public education and awareness, product labels, carbon content labels

---

2050. The state is legally bound to reach carbon-neutral electricity by 2045. The California Energy Commission (CEC) describes this as "a landmark policy requiring renewable energy and zero-carbon resources to supply 100% of electric sales to end-use customers by 2045."[39] It is a learning process: the three lead agencies (CEC, the California Air Resources Board, and the Public Utility Commission) regularly assess and report on the state of technology, transmission capacities, affordability, local reliability, and barriers to and benefits of the program.

California's 2045 goal reinforces a long history of energy innovation in the state. A landmark was the 2006 Global Warming Solutions Act, which capped greenhouse gas emissions at 1990 levels. General goals are backed up by specific 2030 targets: 50% use of renewables, a 50% cut in vehicle oil use, and a doubling of efficiency in existing buildings. The emissions trading system serves as a backstop for other tools: fuel economy/emission standards; the ZEV rules; transport planning; integrated electricity resource planning; and coordinated water and energy efficiency planning.[40]

Also discussed in Chapter 5, Denmark is a renewable energy leader for its record with wind. It shows how domestic resources and smart politics may advance clean energy. Experts attribute Denmark's success to several factors.[41] One is a history of turbine development and technology commitment from government. The turbine industry dates back to the early 1900s, with a revival in the 1970s focused on technical quality. Denmark built on these advantages with economic support, including feed-in tariffs and subsidies. Community ownership built up local support. As Niels Meyer notes, "most Danish turbines are owned by private households based on neighborhood cooperatives."[42] Denmark considered nuclear in the 1970s, but opposition from activist groups and others led its parliament to rule that source out in 1985.[43]

Spain offers a cautionary tale on the transition to wind and solar. Endowed with ample wind and sun resources, Spain moved to capitalize on this advantage in the 1990s by adopting ambitious targets, feed-in tariffs, and other economic support for wind and solar developers. Generous feed-in tariffs were put in place for wind in 1997 and solar in 2004. They worked. Wind capacity grew forty-fold in twelve years. In 2008, Spain accounted for almost half of new global PV capacity. This forced grid innovation: "Out of necessity, Spain pioneered the integration of large amounts of renewable energy into the grid."[44] Yet the government overcommitted. It was obligated to make up the gap for utilities between revenues and costs, the "tariff deficit."

When the financial crisis hit in 2008, electricity demand fell, and the government could not afford the subsidies. A center-right government retroactively cut them. Renewable energy, which "for years was nourished and pampered," became "an unwanted and costly extravagance."[45] By 2013, the head of Spain's renewable energy association said the industry had "gone from misery to ruin."[46] With ample wind and sun, land for development, geographic isolation from other grids, and pressure from EU goals, a rapid transition made sense. Still, Spain could not "bring about a genuine and resilient transformation … similar to Germany's *energiewende*."[47] Flawed policy tools enabled firms with fossil fuel assets to force the government to withdraw its support for renewables. Policies should be predictable and durable. As one industry official put it: "Spain used to be a leader. Now the problem with Spain is that you don't know what they will do after tomorrow."[48]

Kenya offers another perspective on the energy transition. While affluent economies have mature electricity systems and universal access, developing countries seek to expand access. The goal is less to undo carbon lock-in than to minimize it by building a clean energy system from the ground up. Africa will shape the future of energy. Half the people added to the world population by 2040 will be African, and the continent should experience high growth rates in this century.

The IEA projects that Kenya's economy could be four to six times larger in 2040 than it was in 2018, with a GDP of a trillion dollars.[49] Kenya presents a case for building a clean energy system virtually from the ground up. It has geothermal resources that could generate much of its electricity by 2040, with plans for major wind, solar, and hydro growth (although water shortages constrain hydro). It has made major progress in electricity access, which grew from a mere 20% in 2013 to nearly 85% in 2019.[50] Kenya ranks among the top ten countries in geothermal, is home to Africa's largest wind facility (Lake Turkana Wind Farm), and is expanding its solar capacity.[51]

Kenya depends on foreign investment and adopts policies to lower financial risks. Among its advantages are its comprehensive and stable policy framework (a 2019 Energy Act and National Electricity Strategy), investment in transmission capacity, and use of multiple policy tools. Kenya created a feed-in tariff in 2008 on a twenty-year time frame; it applies incentives (exemption from taxes and import fees), uses net metering to promote expansion of distributed solar, and invests in renewables.[52]

California illustrates the use of multiple tools. Denmark shows how to use available resources and make clean energy appealing. Spain highlights the risk of moving too fast and overcommitting. Kenya demonstrates the links between economic growth and clean energy strategies.

Explaining Clean Energy Policy Adoption

Why do some places move faster on clean energy than others? Researchers have studied many factors: institutional character- istics; regime types; economic composition; relationships with other countries; multilateral engagement; public opinion; and political platforms and policies.

One study is highlighted here to illustrate how policy scholars have explored these factors. It uses a framework based on interests, institutions, and ideas. *Interests* refers to the economic, environmental, and other groups having influence; *institutions* to the organization and functions of nations, institutional characteristics, and quality of governance; and *ideas* to the "underlying worldviews and ideologies" of leaders, activists, and the public. Also assessed is *international influence*, which refers to "the economic, political, and strategic relations between countries."[53]

With respect to interests, a constraint is having a large fossil fuel industry and economic dependence on it. Fossil fuel and related industries (vehicles, petrochemicals) can block progress. Workers and consumers perceive their well-being – jobs, low energy prices – as depending on the coal, oil, and natural gas sectors. This is why the clean energy transition progressed more quickly in the EU than in countries like the US and Australia, where fossil fuel industries hold political power.[54] Even in Europe, countries with large coal sectors have difficulty moving beyond fossil fuels.

Of course, these dynamics can change. Prospects improve when interest groups promoting clean energy gain leverage. As renewable industries take on economic importance and gain in political power, their influence grows. The growth of renewable sources around the world creates new political as well as energy dynamics. A well-known public policy collection in the 1990s was titled *Do Institutions Matter?*[55] They do, but their influence on policies is difficult to assess.

Ideas encompass many attitudes, beliefs, and influences. The authors of the study use climate change awareness and levels of social trust to explain progress or a lack of it. They predict that the higher the awareness of climate change and its impacts, the more likely a country will make progress. In turn, low levels of social trust make collective action difficult; low trust also may lead to a populism that is hostile to clean energy policy. Among the other findings of this study are that democracies low in corruption and high in climate awareness are unlikely to maintain fossil fuel subsidies; countries with weak institutions and low climate awareness are less likely to pursue clean energy; and countries having strong institutions (effective administration, low corruption) are more likely to adopt clean energy laws and cut fossil fuel subsidies. Still, many countries with strong institutions, high trust, and climate awareness have high emissions, suggesting that there is "a well-differentiated subset of fossil-extractive nations in which emissions are very high."[56] This may explain the low rankings of the US and Australia in assessments like the *Climate Change Performance Index.*

The study identifies five country clusters (see Table 8.3). Constraints on clean energy are highest in the top row and decline as we move down the rows. These are hardly encouraging results for a transition. Countries dependent on oil, gas, and coal, along with fragile democracies, account for two-thirds of emissions in the countries studied. Even the bottom cluster has laggards as well as leaders like Sweden and Germany. Many studies find EU membership is correlated with strong environmental policy.[57]

This study suggests that some of our assumptions about factors supporting clean energy policies should be modified. It concludes that good governance, low corruption, and high social trust promote clean energy policies. A large fossil fuel sector is a constraint. Several developed as well as transitional economies depend on fossil fuels. Finally, although affluence historically has led to higher emissions, some wealthy countries – examples are in northern Europe – adopt strong policies.

Policies for Accelerating the Clean Energy Transition

Should a country maximize its investments in technology research or adopt feed-in tariffs? Should an economy-wide carbon tax be preferred over renewables quotas? With carbon pricing, does a tax or market creation make more sense? Should clean energy advocates give priority to carbon pricing, renewable quotas, a national energy efficiency standard, or EV tax incentives? The temptation is to say *adopt all of the above.* But the world is too complicated for easy answers and the cost of failure too high. This part of the chapter sets out some lessons for accelerating a transition.

Policy learning is essential

There is not a lot of room for error. Making the wrong choices or foreclosing options that are needed later could cost decades of progress. Understanding what works and does not, and having the

Table 8.3: Country Clusters and Clean Energy Progress

Cluster	Country Characteristics	Country Examples	Percentage of Global Emissions	Percentage of Global Population
Strong dependence on oil & gas extraction	• Lowest adoption • Highest subsidies • Low democratic norms • High corruption	Saudi Arabia Iran Venezuela Algeria	6%	5%
Fragile states	• High corruption • Low social trust • Low democratic norms	Nigeria Pakistan Mexico Bangladesh Kenya	3%	11%
Heavy coal reliance	• Much rapid growth • Mixed regime types • Some high corruption • Retain subsidies	China India Turkey South Africa Russia	47%	48%
Fractured democracies	• Electoral democracies • Some corruption • Low trust • High for policy; low subsidies & emissions	Brazil Argentina Chile Greece Hungary	4%	6%
Wealthy OECD countries	• Moderate trust • High awareness • Low subsidies • High emissions • Affluent	France Sweden Italy Spain Germany	30%	13%

Note: Percentage of emissions and population do not total 100% because some countries were not included in the clusters.
Source: Based on William F. Lamb and Jan C. Minx, "The Political Economy of Climate Policy," *Energy Resources and Social Science* 64 (2020).

right policy tools for use in different contexts, require policy learning within and among countries.

Countries draw upon the experience of others. Research suggests that policy tools diffuse in three ways: *emulation*, when there is information sharing and learning; *coercion*, when policies are adopted due to a forcing mechanism, such as an EU directive or multilateral aid; and *competition*, when countries monitor what others are doing and adopt policies for competitive advantage. One study found that "countries are significantly more likely to adopt renewable energy policies used by countries with similar political and energy sector conditions."[58] But the specifics varied. With feed-in tariffs, countries tend to follow the lead of their political and economic counterparts; with quotas, they follow countries that have used them successfully. FITs are seen as riskier and more costly, the study found. Countries receiving foreign aid tend to learn from and copy their donor countries.

Choices among policy tools are also informed by effectiveness studies. For example, we have evidence that efficiency policies can make a difference. In an effort to assess overall policy effectiveness, researchers did a "meta-analysis" of energy efficiency research. They found efficiency policies to be effective in reducing energy demand, and that a "combination of instruments generates a significantly larger impact on energy demand" than any single one, confirming the need to leverage various tools in strategies.[59] They recommend a focus on residential and commercial sectors, where policies have the most impact, and understanding how tools may be used in combination.

Policy experts are learning more about effectiveness. Carbon pricing moves countries away from coal and toward nuclear and renewables.[60] One quantitative analysis of data from 142 countries found that the annual growth rates in CO_2 emissions from fossil fuel combustion were some 2% lower in jurisdictions having a carbon price; an additional 10 euros per ton tax was associated with a 3% slower annual rate of emissions growth.[61] Policies can also have spillover effects. One study found that "countries that price carbon emissions have gone on to adopt more wind and solar energy."[62]

Mandates like an RPS or a FIT are found to account for renewables expansion in places like Germany, Spain, Denmark, and some US states. Less agreement exists on how to design policies that not only work but minimize costs in achieving an equitable transition.

Studies suggest that effective strategies combine the strengths of different tools. There is evidence from the US, for example, that a state RPS combined with the federal Production Tax Credit boosted wind energy.[63] The RPS motivated utilities to expand wind generation, while the PTC lowered investment risks and stimulated financial commitment. Carbon pricing discourages the use of fuels or practices that emit carbon.[64] Combined with an RPS or FIT, carbon pricing promotes renewables investment. Policymakers can use ZEV mandates or gas/diesel bans along with consumer incentives to promote EVs. Affordable financing may be combined with energy audits, building codes, and tax incentives to cut energy use and intensity. Fortunately, there is a rich body of policy research on what works for clean energy. This discussion highlights some of the leading work.

Solutions vary for countries and regions

What works in Spain or Kenya may not work in India or Brazil. Countries vary greatly, starting with their resources and endowments. Wind works in Denmark because it has experience with the technology, enjoys ample wind resources, and ensured local support. India has plenty of sun and so seeks to electrify rural areas with distributed PVs. Offshore wind has natural advantages for places like Scotland and Ireland. Political choices also matter. For better or worse, Germany took nuclear off the table. Powerful fossil fuel sectors make effecting change hard, especially in Russia and Venezuela. What is feasible in a centralized system like China or France may be harder in federal Argentina.

Economic structure matters. Mature economies are working, to varying degrees, on turning fossil fuel societies into ones built on renewables. Rapidly growing countries like China and India struggle to meet huge energy appetites with what they have, which is lots of coal for electricity and industry, while expanding renewables and improving efficiency. Countries in Sub-Saharan Africa confront other choices. Some economies have a manufacturing industry, while others rely on services, tourism, or resource extraction. Although decarbonization includes all the pillars discussed in this book, the specifics will differ.

Political feasibility also varies. Carbon taxes are common in Europe, even with the EU's trading system, but have proven nearly impossible in the US. Government's role in the economy is more accepted in northern Europe. It is harder to give up carbon when economies are built on fossil fuels. The politics within individual countries also matters: of the bottom twenty states in the ACEEE's *State Energy Efficiency Policy Scorecard*, nearly all are politically conservative. Indeed, research suggests that political attitudes affect views on seemingly non-ideological issues like energy efficiency.[65]

Politics affects the feasibility of policy options. Taxes particularly face an uphill climb. It is no accident that half the American states have adopted an RPS, while none have enacted a carbon tax. This is despite evidence that a tax would deliver less costly emissions reductions. It appears that, in the US, carbon pricing in the form of cap-and-trade is more feasible than an outright tax. The state of Washington tried three times to adopt a carbon tax, emulating its Canadian neighbor, British Columbia. Two such efforts were as voter initiatives, and the third as legislation. All failed. Finally, in April 2021, the state enacted a cap-and-trade program that may someday link up with California's.[66]

Why did cap-and-trade succeed and a tax fail? Ideological resistance to taxes played a role, as did concern about costs. Voters had little faith in promises that funds would be distributed back to households. Well-funded opposition doomed the tax, while cap-and-trade faced less pushback. One study found that Vermont is the only US state likely to adopt a tax in the near future.[67]

How issues are framed makes a difference. Surveys find that climate change is the most divisive issue in US politics.[68] Still, conservative states (Iowa, Wyoming, Texas) lead in wind. In these states, it helps to frame the issue as one of access to low-cost, reliable, clean energy rather than as a way to fight climate change, a framing that pushes people into their ideological corners.

All the parts have to be coordinated

In discussing clean energy with students, I often find they assume it is all a matter of scaling up renewables. Renewables are critical, but clean energy is about more than scaling up wind and solar. It is about efficiency; transport, industry, and buildings; integrating renewables; a future with hydrogen and CCS; and more. A theme of this book is that the various parts of national and global energy systems are inter-related – that no one part can be isolated from the others. If energy demand grows as fast as renewables, it is not much of a transition. If transmission capacities do not expand, or if EVs are powered by coal-generated electricity, little is gained.

Our understanding of the effectiveness of various tools reinforces the value of a system-wide approach. This starts by keeping efficiency in the forefront. Studies of RPS in the US find that they are effective in increasing the amount of renewable electricity, but less effective in changing the generation mix.[69] Renewables chase a moving target if demand keeps growing. Infrastructure, transmission, and technologies need to be aligned with the goal of integrating lots of renewables.

Stable and diverse policies help

Investors, businesses, and consumers need predictable signals. Uncertainty is an enemy of innovation, especially in areas as complex and multifaceted as energy. Policy research finds that technology innovation is associated with *stringent* regulatory standards, *flexibility* in meeting them, and *predictability*.[70] For example, carbon taxes should include increasing but predictable prices over time so investors and businesses can plan. Erratic shifts undermine their effectiveness.

Mandates illustrate the value of stability and predictability. Hawaii and California are clear in their expectations for 100% renewable and net-zero-carbon economies by 2045. Consumers know fossil fuel cars will not be sold in Norway in 2025 and will not be legal in Oslo anyway. Financial incentives especially should be predictable. Lapses in the US Production Tax Credit are associated with declining wind deployment.[71] Change is hard when "you don't know what they will do after tomorrow," as the Spanish official put it.

Diverse ecosystems are healthier and more resilient. Financial advisors recommend diverse portfolios. Does the same principle apply to policies? Three scholars say that it does. They develop a measure of policy tool design, the Average Instrument Diversity (AID), to assess the relationship between a diverse, tailored policy portfolio and effectiveness. They find that tailoring policy tools to specific problems and contexts makes them more effective. The analysis examines the policies of twenty-one OECD countries from 1976 to 2005 on three issues: clean air, water protection, and nature conservation. It assesses such tools as regulatory mandates, economic incentives, and information disclosure. Countries vary in their policy diversity – the extent to which tools and strategies are tailored to particular problems. The study concludes that the more diversified and tailored a strategy, "the higher the chance that the policy design takes account of the nature of the underlying policy problems."[72]

Policies should be just and equitable

There are both pragmatic and ethical sides to an energy transition. The practical side is to make it work – to have a largely or entirely carbon-free energy system by the middle of the century. That in itself represents a gain in equity because most social costs fall disproportionately on the poor and vulnerable. Climate instability affects everyone, but especially those lacking the resources to respond or who depend on ecosystems. Health damages fall heavily on urban areas, especially in high-growth areas where air quality is worst. Damaged forests and landscapes hit the poor the hardest.

The energy transition itself adds to equity, but less so if it is done unfairly. If energy costs go up for the poor, if the transition leaves fossil fuel workers behind, or if access to family-supporting jobs is distributed inequitably, then it is not a **just transition**.[73] Three brief examples make the point. By design, carbon pricing makes energy cost more. Low-income groups spend a higher proportion of their disposable income on energy. If revenues are not cycled back to them, their lot is worse. Coal communities are built on a commodity that is being phased out. If governments do not help, these communities are worse off. Tax incentives promote efficiency that reduces costs, but low-income groups may not be able to benefit from them. In each of these cases, the transition is unjust.

Public Policy, Government, and the Energy Transition

Government played a role in previous energy transitions – creating legal relationships, granting subsidies, building infrastructure, funding research, and issuing standards. This time, it must play a more active and intentional role: guiding, coordinating, and funding the change. Markets are sound tools but they have to be guided, enhanced, and balanced. The next chapter considers the prospects for success and fairness in the energy transition. Getting the carbon out requires global action, technology and social innovation, political commitment, lots of financing, and a good bit of luck.

Guide to Further Reading

Alex Bowen and Sam Fankhauser, "Good Practice in Low Carbon Policy," in Alina Averchenkova, Sam Fankhauser, and Michal Nachmany, eds. *Trends in Climate Change Legislation*, Edward Elgar, 2017, 123–42.

Sanya Carley and Michelle Graff, "A Just U.S. Energy Transition, in *Handbook of U.S. Environmental Policy*, ed. David M. Konisky. Edward Elgar, 2020.

Barry G. Rabe, *Can We Price Carbon?* MIT Press, 2018.

World Bank, *State and Trends of Carbon Pricing*, 2021.

9

The Clean Energy Future

The last eight chapters examined the need for clean energy and the forms it may take. The need for a transition is not up for debate. Greenhouse gases accumulating in the atmosphere, most of them related to the global energy system, make decarbonization an imperative. Climate change combines with air pollution and ecological degradation to make a compelling clean energy case.

Yet a transition does more than deliver us from harm: it also presents opportunities. Using energy efficiently saves money, frees up resources for things like education and health care, helps make energy affordable for low-income groups, and opens up productive avenues of employment. Renewable technologies create jobs; their distributed and decentralized aspects expand options for communities and households to control energy resources. Energy efficiency frees up resources for meeting other social needs and advancing equity. Distributed energy enhances security, expands electricity access, and supports community control over resources. Advanced technologies and smart grids open up new markets, research paths, and careers to people around the world.

My concern in this chapter is with the prospects for clean energy. The transition has both technological and social aspects, and both matter. There is also the political aspect – how the clean energy transition is framed and the conditions in which it will occur. The barriers to change lie less in technology and economics than in politics. How do we punctuate the carbon policy equilibrium?

Technology Innovation and the Energy Transition

Technology is viewed from many perspectives in clean energy discussions. One view is that technology created the problems we are urgently trying to deal with. Carbon-based energy systems were built on nearly two centuries of technological change. Nuclear power is about as complicated a technology as one can imagine. Electricity grids powered by fossil fuels are marvels of modern engineering. ICEs use a liquid fuel that transformed twentieth-century life. These technologies and their impacts may be seen as problems now, but all were hailed in their time as pathways to a better world.

For our purposes, technology is considered a neutral tool. It is not the only relevant tool, to be sure, but it is a big one. Later I consider the social aspects of the energy transition: energy justice and democracy. Those matter, but technology innovation does as well. The reality is that we will not get anywhere close to a decarbonized world if a great deal of technology innovation is not scaled up. Short of a catastrophe, the world will not return to a pre-industrial past. Energy is too wrapped up in modern life – in mobility, heating, cooling, manufacturing, communications, entertainment, health protection, and much more – for us to live without it.

Stages in clean energy innovation

It should be clear by now that inventing technologies, even demonstrating them, is not enough.[1] Energy's history is full of promising technologies that never advanced beyond laboratory or pilot phases. It is worth looking at the innovation process to understand what needs to happen.

Energy technology innovation is "the process of generating ideas for new products or production processes and guiding their development all the way from the lab to their mainstream diffusion into the market."[2] Having an idea or coming up with an invention is only the start. The IEA sees innovation as occurring in four stages: prototype, demonstration, early adoption, and maturity. *Prototypes* translate ideas into designs, into examples of a product or process that could work. A *demonstration* is a full-scale commercial application that shows it can serve an intended use. In *early adoption* a product or process is scaled up, commercialized, and gains competitive status in markets. This is where most innovations falter – in the valley of death. A technology may struggle to close cost and performance gaps against its competitors, usually incumbents that are locked in. Finally, *mature* technologies become established in markets, have cost and performance parity with incumbents, and enjoy commercial success.

A few illustrations will help. Wave energy generates a fraction of a percent of world electricity. Decarbonization scenarios see that growing slightly by 2050. Despite its potential, wave energy has a "forever round-the-corner" reputation as a good idea that does not scale up and is stuck at demonstration stage. But there are hopes. A 2020 analysis speculated that wave energy may be "emerging from the trough of disillusionment and could soon be ascending the slope of enlightenment."[3] For now, wave technologies struggle to get through the valley of death. BECCS is another technology that is unproven commercially. Until CCS is proven at scale, BECCS is unlikely to be competitive. In the last decade, PV has achieved mature technology status; along with onshore wind, it now leads in newly added annual generating capacity in most of the world. PV and onshore wind have closed performance and cost gaps with fossil fuels; their clean energy attributes at this point are frosting on the cake.

In a revealing article, Robert Fri recounts how the US Department of Energy spent $13 billion on technology innovation from 1978 to 2000. This generated $40 billion in economic benefits – a highly favorable return.[4] Yet a mere *0.1% of the $13 billion invested generated three-fourths of the benefits*, and they came from three technologies: high-efficiency refrigerators, electronic ballasts for fluorescent lighting, and low-emissivity windows with coating that improves efficiency. This pattern is not unique to energy. Investments fail; that is part of the process. Innovation is incremental and cumulative, requiring patience and long-term thinking.

Drawing on a US National Academy of Sciences project that he led, Fri offers advice that remains relevant. Most technology innovation occurs in the private sector; government's primary role is to let private capital carry the load. Public policy should push innovation with the tools examined in Chapter 8 and by easing the path to commercialization. Fri prefers economic and regulatory tools to subsidies, which are risky, as the US experience with the solar module company Solyndra shows.[5] Government should lead basic research where private incentives are lacking, but otherwise "set the table" for innovation. It should do applied research to speed up innovation, as well as create markets, such as through purchasing policies for products like EVs and solar panels.

Energy technology innovation occurs in a complex environment, one with a great deal of long-lasting, capital-intensive physical infrastructure, defined by intricate relationships among the parts of the system.[6] Government plays a necessary role in technology innovation, but we should heed Fri's advice that it cannot fund all of what is needed, and that policies should be designed to motivate the private sector to do much of the work.

The state of clean energy technology

The IEA's 2020 report *Accelerating Technology Progress* classified 400 technologies by their stage in the innovation process.[7] These technologies are called for in the IEA's Sustainable Development Scenario, which defines a path to meeting the Paris goals. Of the 400 technologies, 17% were rated at *prototype* stage; 17% at *demonstration*; 41% at *early adoption*; and 25% as *mature* technologies. Thus one-third of the technologies needed to meet the Paris goals fall short of early adoption or commercialization.

In another assessment, the IEA rated forty-six "critical" *categories* of clean energy technology to determine which are on track for mid-century.[8] Of these, five were rated *on track*, or at the stage of commercialization where they should be; two involved buildings and efficiency (lighting, and data centers and networks); two were in transport (EVs and rail technology); and the other was PV. Ten were rated as *off track*: nuclear, CCUS in power and industry, transport biofuels, building structures, and heating. The rest were rated as *more effort needed*. These included renewables, natural gas, aviation and shipping, appliance efficiency, heat pumps, storage, smart grids, and hydrogen.

Tables 9.1 and 9.2 present some of these categories. Table 9.1 lists those for electricity. Table 9.2 lists those for energy integration, transport, and industry. The comments are the IEA's views on what should happen by 2030 to meet the Paris goals: they illustrate the progress that is needed to expand renewables, transform transport and industry, and integrate renewables into electricity.

Table 9.1: Technology Development/Deployment: Electricity Generation

Technology Area	IEA Assessment	Comments on Needs
Solar PV	On Track	Strong recent growth but needs average annual growth rate of 15% by 2030
Onshore Wind	More Effort	Steady growth but needs to scale up the rate of additions to generating capacity
Offshore Wind	More Effort	Progress in Europe; needs to extend that progress in cost reductions, technology, & deployment
Hydro	More Effort	Needs to have annual average growth rates of 3% by 2030
Geothermal	Not on Track	Needs to grow by an annual average rate of 10% by 2030
Ocean	Not on Track	Now at a very small base; needs to have annual average growth rate of 23% by 2030
Nuclear	Not on Track	New construction is lagging; needs to extend lifetimes of existing plants & double the annual rate of additions
Natural Gas	More Effort	Use without CCUS should disappear at some point; needs work on commercializing CCUS
CCUS in Electricity	Not on Track	Well off track; not now commercially viable; needs targeted policies, especially financial incentives

Source: IEA, *Tracking Clean Energy Progress*, 2020.

What would my list of technology needs include? Certainly, there is work to do on the next generation of PV. Although silicon-based PV has become a mature technology, experts like Varun Sivaram warn about being too complacent and thinking only incrementally; to stay competitive, especially given PV's variability, alternative materials and technologies should be pursued.[9] Even a mature technology like wind offers low cost/high efficiency potential with vertical axis turbines that are more productive and use less land than the now-dominant horizontal axis designs. Battery storage is another

Table 9.2: Technology Development/Deployment: Systems Integration, Transport, Industry

Technology Area	IEA Assessment	Comments on Needs
Energy Storage	*More Effort*	Double-digit growth is required to meet electricity needs; progress depends on financial & market creation policies
Hydrogen	*More Effort*	Needs technology gains and lower costs for electrolysis (green hydrogen) and needs to expand applications
Smart Grids	*More Effort*	Steadily more decentralization/digitization; needs more use of technologies like AI; more policy support
Electric Vehicles	*On Track*	Many signs of progress; continue rapid growth and policy support; link to decarbonized electricity
Passenger Vehicle Efficiency	*Not on Track*	Incremental gains, but the pace has to be greatly accelerated
Aviation	*More Effort*	Most progress is in efficiency; now a need for new designs and better low- and zero-carbon fuels
Ocean Shipping	*More Effort*	Need for greater efficiency as well as low- and zero-carbon fuels and technologies
Iron & Steel	*More Effort*	Need for reducing carbon intensity further with research/deployment support and market creation
Cement	*More Effort*	Needs technology gains and support for CCUS in industry
CCUS in Industry	*Not on Track*	Only technology for decarbonizing some industries; needs policy for lower costs, simpler applications

Source: IEA, *Tracking Clean Energy Progress*, 2020.

area for avoiding technology lock-in. The now-dominant lithium-ion designs are limited in their capacity for large-scale, long-term storage.[10] Another priority is green hydrogen. It enhances storage capacities and has transport applications. Hydrogen fuel cells can offer solutions for heavy-duty transport and ocean shipping, as well as for hard-to-decarbonize industry sectors.

What about those probably necessary but controversial technologies: advanced nuclear and CCS? Given our interest in keeping all feasible options on the table, there are good reasons to develop and scale up a next generation of nuclear, especially small, modular reactors. If issues of cost, safety, and use flexibility are overcome – an admittedly tall order – advanced nuclear has a role. CCS and its variations (CCUS and BECCS) raise difficult issues. Even removing more than 90% of the CO_2 means that some is released to the atmosphere, along with particulates and other local pollutants. Yet both help with tough industry sectors and limit harm from remaining coal plants. BECCS could anchor a carbon removal strategy. Technology scale-up still is necessary.

The Social Aspects of the Energy Transition

Technology innovation is critical to the energy transition. Yet technology is part of complex social, economic, and political relationships. Energy affects societies, and social conditions shape energy systems. This section examines the fairness, political feasibility, and social effects of clean energy. One set of topics is *energy justice* and a *just transition*. Another is *energy democracy*. Energy injustice exists when people "do not have access to safe, affordable, and sustainable energy, are not involved in decisions related to energy access, or are disproportionately burdened by changes in the energy system."[11]

The just transition: enhancing justice and equity

This is the first energy transition in history being driven largely by social costs. In addition to pumping CO_2 into the atmosphere, carbon-based energy causes formidable environmental and health damage. If done fairly, moving to a clean energy system helps vulnerable groups in society. Better air and water, healthier ecosystems, and a stable climate system make everyone better off.

This is not inevitable, however. Even clean energy can leave some groups, within countries and across regions, getting the short end of the stick. If a new energy system increases economic and social inequality by making energy less affordable, fails to account for its effects on displaced workers and communities, neglects paths for expanding access to jobs and other benefits, or uses policy tools and strategies that increase rather than ameliorate economic and social inequity, then it is unfair. A just transition will require changes in the energy system that do not impose excessive burdens on some in society while granting too many benefits to others.

One path to a just transition is to give attention to energy affordability. Access to energy services is a *merit good*, defined as: "goods and services that should be accessible by all individuals in society irrespective of their ability to pay."[12] Energy is a good that people need to sustain a quality of life, and should be widely available.

A US study found that low-income households (those below 80% of median income) spend over 7% of their earnings on energy, while high-income households spend just above 2%.[13] Energy burdens were more than three times higher for low-income households. Some studies find an even higher low-income burden, up to 20%.[14] The highest burdens in the US are in South Carolina, Mississippi, and West Virginia, all ranking near the bottom in the ACEEE's energy efficiency policy ratings. States with the most to gain from efficiency do the least about it, at least for low-income groups.

These burdens reflect more than just the income shares for energy. For many reasons, low-income groups may not share in the benefits of a clean energy transition. They lack funds to invest in efficient appliances and lighting; they are renters who cannot afford upgrades like installing insulation; they drive older, inefficient cars. As we have seen, efficiency on its own is a path to lower energy burdens. Upgrades like state-of-the-art insulation and high-efficiency appliances can cut consumption significantly.

Even progressive jurisdictions like California still have to account for unfairness in policy. A study by the University of California at Los Angeles found that because the poor use less energy and pay lower taxes, they benefit less from many incentive programs. Tax credits for EVs, efficiency, and residential solar mostly benefit high earners with high marginal income tax rates. Indeed, UCLA found that such incentives "are disproportionately benefiting wealthier homes that use more energy than they need to live comfortably."[15]

Also illustrating a just transition are effects on displaced communities and workers. The overall effects of clean energy will be positive, but that does not mean some are not harmed.[16] Among those bearing burdens are workers in the fossil fuel and related industries and the communities where they live – coal miners, oil workers, others in carbon-dependent sectors. Some sectors will endure but change. EV manufacturing is less labor-intensive than for ICEs, product life spans are longer, and maintenance is easier, all with job impacts. The transition creates job opportunities in renewables, efficiency, grid modernization, mass transit, and infrastructure, but other sectors will suffer. For regional economies based on fossil fuels – coal areas in Poland, eastern Germany, and the US Appalachia – a transition threatens the economic and social viability not only of workers but of entire communities.[17]

Why should governments worry about a just transition? To start with, it is the right thing to do, especially given government's role in accelerating decarbonization. Public policy adds to losses suffered by vulnerable workers and communities. Second, a just transition helps maintain a decent quality of life, which should be what government aims to deliver. Having displaced workers in declining communities with poor education and health care and an array of social problems does not reflect well on any of us. Third, supporting workers and communities helps build a clean energy coalition and mitigate political opposition.[18] Fear of economic displacement generates support for parties opposing clean energy and, for that matter, democracy. Eastern Germany's coal regions are fertile ground for the right-wing Alternative for Germany.[19] In the US, coal areas helped Donald Trump win in 2016, an outcome that had profound effects on energy and climate policy.

This issue draws attention in Europe. The EU's Green Deal includes a Just Transition Mechanism (JTM) that is focused on "regions that are the most carbon-intensive or with the most people working in fossil fuels."[20] The JTM aims to mobilize at least 150 billion euros in investment from 2021 to 2027 to promote job options, retrain displaced workers, support energy-efficient housing, expand clean energy access, and reduce energy poverty. At national and regional levels, the JTM seeks to create clean energy jobs, invest in public transport, enhance digital connectivity, and improve infrastructure – a formidable social agenda. The practical and ethical basis is made clear in a statement by Frans Timmerman, Executive Vice-President of the European Commission: "We must show solidarity with the most affected regions in Europe, such as coal mining regions and others, to make sure the Green Deal gets everyone's full support and has a chance to become a reality."[21]

There has been less of a focus on displacement in the United States. An illustration of what may be done is Illinois Senator Tammy Duckworth's proposed Marshall Plan for Coal Country Act of 2020. Reflecting the goals of a just transition, it would make "investments in communities that have powered the United States and ensure that these communities are able to prosper and benefit from the energy transition."[22] The bill would provide health care for displaced workers, college tuition, a minimum wage, protection for worker benefits, environmental restoration, and incentives for local economic development.

Although coal workers have suffered most so far, the transition also affects other industries. A 2018 report by the International Labor Organization (ILO), comparing business as usual to a clean energy scenario, estimates some 6 million job losses in petroleum refining, oil and gas extraction, coal and lignite mining, and electricity production from coal, among others. At the same time, the ILO forecasts some 24 million added jobs in construction and electricity production from renewable sources.[23] Given their role in accelerating clean energy, governments should devise policies that equitably manage such impacts and provide for a just transition.

Energy democracy: transforming politics with technology

Do the ways in which energy is produced, distributed, and consumed in society reflect or influence distributions of and access to political power? Writers on energy democracy say they do.

Energy democracy grew out of other energy-related movements, especially anti-nuclear and climate justice.[24] It views the energy transition as being about more than technology, or renewables replacing fossil fuels. It should be a social, political, and economic process of transforming the ownership and control of resources and models for allocating benefits and costs. Its theme is that "energy systems are inseparable from larger social and ecological paths and relationships" in a society.[25]

From this point of view, energy's history is one of concentrating political power and control in large, centralized, technologically complex systems that deliver benefits to powerful interests while pushing the social costs onto others. Nuclear is a classic case, with large facilities owned by corporate interests and regulated by distant government. Indeed, the historical opposition to nuclear by European green parties is based as much on social attitudes as on pragmatic perceptions of safety. Fossil fuels also fit this historical pattern of generating social costs: pollution in vulnerable communities; oil refineries in low-income areas; destroyed ecosystems due to fossil fuel production; energy burdens for the poor; economic dependence on big, distant utilities or producers with scant regard for those affected by their decisions.

For energy democracy advocates, fossil fuels, especially given the impacts of climate change, reflect patterns of concentrated economic and political power, gross inequities in the distribution of social costs, and insulation from community and citizen control. International oil companies are a target. Until the rise of the technology industry, oil companies were the world's largest for-profits. Oil firms withheld information about climate risks, further undermining their legitimacy.[26] One path to democratized energy is community cooperatives, discussed in Box 9.1.

Electrical grids and the political and economic power of big utilities also offer a ripe target. From their fairly primitive urban origins late

---

Box 9.1: Energy Cooperatives
Energy cooperatives are owned and governed by their members and enable community control of energy resources. They focus especially on renewables and energy efficiency. Rooted in the principles of "self-help, self-responsibility, democracy, equality, equity, and solidarity," their features include voluntary membership, direct ownership of resources, participation in pricing, cooperation with governments, and environmental concern.[27] The distributed form of renewable resources makes cooperatives possible. They are well developed in Denmark and Germany, but are expanding in many countries. Denmark has a tradition of collective ownership of entities like banks and dairies. It built on this tradition in its wind expansion, where community ownership and participation promoted acceptance. In one community wind development, at Middelgrunden near Copenhagen, the community was involved in the placement of turbines, and receives 50% of revenues. The wind farm was "co-funded, co-managed, and co-owned by the people of Denmark" and the community.[28]

in the nineteenth century, electricity grids were consolidated and integrated in the twentieth. There were good reasons for this at the time, given the dominant carbon-based technologies and the need to integrate the physical infrastructure making up modern grids, but this further concentrated economic and political power. Energy systems are defined by complex technologies, substantial physical infrastructure, access to large financial capital, and formidable corporate organizations distant from the people using the energy. For energy democracy advocates, the opportunity here is to shift from a system that benefits centralized, corporate, utility-scale owners to one that is accessible to communities and designed to serve their needs – in other words, to confront and destabilize the "dominant systems of energy power" and move to "decentralized, democratized, and community-based renewable energy futures."[29]

Renewables, especially PV, are scalable. In contrast with fossil fuel technologies that require big facilities, complex supply chains, lots of capital, and that often concentrate power, renewable technologies are *distributed energy resources*. They are ideal for community-based and household ownership. Microgrids enable community control and ownership. Two-way communication with smart grids further enhances prospects for local control.

Will the energy transition perpetuate the patterns of economic and political power created by existing carbon and nuclear systems, or will it build something more democratic, decentralized, equitable, and community-based?[30] Decarbonizing offers at least the potential for more equitable, community-oriented energy. As well as promoting equity by reducing social costs, the transition may help expand access to energy and make it more affordable, although this depends on how it is allocated and priced. For example, carbon pricing helps with nearly all aspects of the transition – it promotes efficiency, discourages fossil fuel investments and use, gives economic advantage to renewables, and generates infrastructure revenue – but it hurts low-income groups. Taxes and market creation should be fair. This can happen by redistributing funds to low-income households, cutting taxes on low earners, and funding efficiency and health care. Box 9.2 describes energy democracy's themes, expected outcomes, and selected initiatives.

Given the state of politics in much of the world, calls for energy justice and democracy are on to something. Political polarization, mistrust, and anti-democratic attitudes are prevalent even in

Box 9.2: Energy Democracy: Policies, Outcomes, and Illustrations

Policies and Practices to Support Energy Democracy:

- On-bill financing (financing efficiency and other improvements through energy bills)
- Community green banks and other ways to increase access to financing
- Support for community-owned microgrids
- Participatory planning in energy resources management; a role in utility governance
- Public benefit programs (reserving energy revenues for local initiatives)
- Carbon pricing designed to redistribute revenues as well as finance clean energy
- Support for energy cooperatives and community energy institutions
- Community choice aggregation support (pooled buying and financing mechanisms)
- Smart metering policies that encourage residential control
- Access to home weatherization and efficiency programs for low-income households
- Long-term, guaranteed energy prices

Expected Outcomes for Energy Democracy:

- Clean energy system with greatly reduced social costs (no fossil fuels)
- More decentralized and distributed economic and political power
- New alliances of social groups
- More social and public control of production and consumption (*prosumers*)
- New ownership and financing models (cooperatives; community-controlled financing)
- Job opportunities for people of color, women, and disadvantaged groups
- Recognition of people as energy citizens, not just consumers (energy as a merit good)

Illustrative Initiatives on Energy Democracy:

• Institute for Local Self-Reliance (US): https://ilsr.org/energy-democracy-in-4-steps
• Rosa Luxembourg Stiftung (Europe): https://www.rosalux.de/fileadmin/rls_uploads/pdfs/sonst_publikationen/strategies_of_energy_democracy_Angel_engl.pdf
• Switched-on London (UK): https://energy-democracy.net/london-united-kingdom

Source: Adapted from Matthew J. Burke and Jennie C. Stephens, "Energy Democracy: Goals and Policy Instruments for Sociotechnical Transitions," *Energy Research and Social Science* 33 (2017), and other sources.

established democracies. Right-wind parties hostile to democracy, climate action, and clean energy hold political power in many places.[31] Although there are political risks involved in broadening the clean energy agenda, as discussed below, there are also political opportunities to build fairer, more just societies and to strengthen democracy. This is a time not only for a just transition but for reducing inequality in "a moment of technological change ... as significant as the computer revolution."[32]

The Politics of an Energy Transition

Implementing fundamental economic and social change is difficult. To a point, this is a positive. Some stability and predictability is an asset. At the same time, systems have to adapt to meeting new needs and realizing new opportunities. Energy transitions historically illustrate adaptation and change. Now we see the need for another energy transition, one that governments will guide and accelerate.

Reaching clean energy lock-in

In his essay on carbon lock-in, Gregory Unruh identifies the barriers to post-carbon energy systems: extensive physical infrastructure, complex economic relationships, the advantage of the familiar, and so on.[33] But there also is a *politics* of carbon lock-in. Politics is a process of reconciling diverse interests in society. Energy transitions require us to do this in new ways. For decades, energy systems have been dominated by coal, oil and gas, nuclear, and allied interests. Such interests may not welcome a transition that, after all, aims to put them out of business.

Under what conditions is a transition to clean energy most likely to occur? It likely will succeed during a period of global economic and political stability. At a national level, we should hope political systems avoid extremes and deliver capable centrist governments. Steady growth helps by promoting political stability, sustaining effective governance, and providing investment funds. Sound multilateral and international institutions are critical, from the EU to the United Nations Environment Program. No country can make a transition on its own. Cooperation, technical assistance, and lesson-sharing are necessary, as argued in Chapter 8.

At some point, if the transition moves along, a *clean energy lock-in* will emerge. An array of fast-charging stations for EVs will replace existing gas and diesel stations. Modernized grids will integrate growing amounts of variable renewables into their generation mix. EV batteries will become part of storage. Utilities will adapt to distributed energy from residential and community sources. Green hydrogen hubs will emerge to support its uses as a source of storage, industrial heat, and fuel cells for mass transit and distance freight. Energy politics will reach a tipping point as clean energy interests acquire political power and are locked in.

There are many paths to decarbonizing, and there will inevitably be debate about precisely what path to follow. What works for Pakistan or Canada may not be the best path for Sweden or China. Nuclear makes sense in some places but not in others. Countries have different histories, resources, lifestyles, cultures, technology strengths, and capabilities that lead to diverse paths Still, there are common elements. This is how the global energy system *could* look by later in this century:

- Coal disappears from electricity generation. This happens first in rich countries, then in rapidly growing ones with newer coal plants. If coal is still needed for specialized uses in carbon-intense industry sectors like iron and steel, it is linked with CCS or BECCS.
- Natural gas remains as a source of electricity generation, almost

entirely for generation at peak demand and other specialized needs, but it is being phased out. Residual emissions from natural gas are captured and stored or used to make products.

- Wind and solar make up 70–80% of global electricity; other renewables (hydro, biomass, geothermal, ocean) contribute 15%, and advanced nuclear accounts for the remainder.
- CCS and CCUS are more economically competitive; although not as cost-effective as other mitigation measures, they will be used to offset hard-to-avoid emissions in heavy industry. BECCS may help with carbon removal. Carbon pricing makes the economics work.
- Passenger fleets are almost fully electric. With falling wind and solar costs, nearly all hydrogen is green. Hydrogen fuel cells are the technology of choice for trucks and buses. Internal combustion vehicles are a relic; oil consumption falls to negligible levels.
- Regional, short-haul air travel is electrified. Longer flights run on advanced biofuels or hydrogen fuel cells. Ocean shipping relies on a combination of advanced biofuels and hydrogen, supplemented by old-fashioned sail power.
- Electricity production and distribution are transformed. Many areas get electricity from distributed energy – small PV, community wind, concentrated solar, microgrids, perhaps wave and tidal – as well as hydro. Bulk grids are sophisticated technologically, smart and interactive, and are linked with EVs. Interactions among grids are smooth and efficient.
- Technology and awareness deliver efficiency in buildings, homes, and transport. Overall energy demand is lower than it was in 2020, and productivity improves dramatically.
- City planners get smart about designs for sustainable land use and efficient mobility.
- Access to clean, affordable, sustainable energy is universal, achieving the UN Sustainable Development Goal, a big victory for global cooperation.

There are many variations in this path, but this is how global decarbonization could look. It almost certainly will build on the pillars of aggressive efficiency, renewables growth (especially wind and solar), and electrification of most transport and industry, with a prominent role for green hydrogen, and selective use of complementary sources like advanced nuclear and CCS/BECCS.

What level of investment is needed to decarbonize? In 2019, the IRENA projected that investment should total a cumulative $110 trillion from 2016 to 2050. This averages to $3.5 trillion per year. Current investment is between 1 and 2 trillion. Of the $110 trillion needed, the largest share in the IRENA projection is for energy efficiency, at $37 trillion (35% of the total); next are renewables at $27 trillion (24%), electricity and infrastructure at $26 trillion (23%), and fossil fuel and related (including CCS/BECCS) at about $20 trillion (18%).[34] If we project the size of the global economy at the halfway point in this projection, say by 2035, this comes to 2–4% of world GDP.

This is a great deal of money but, given the benefits, a sound investment. Climate change costs alone may take a percentage point off world GDP for each degree of temperature rise. Health damages from air pollution amount to 2–5% of global GDP.[35] Even without accounting for the many other ecological damages, putting 2–4% of the world's annual GDP into clean energy is smart investing.

Framing the issues

Albert Einstein once said that "The framing of a problem is often far more essential than its solution." He could have been talking about politics and policy as much as physics. It is well established in the policy field that the definition and framing of issues affect the solutions.[36]

Critics disparage clean energy because they say it harms economies, destroys jobs, and impedes growth. To be sure, it may interfere with growth in certain sectors and eliminate some jobs. But the overall economic and social effects are far more positive than negative. These overall effects are captured in the concept of a *green energy economy* as a way to frame the energy transition.[37]

Political systems focus on the short-term and are usually designed more to protect existing interests than to empower emerging ones. This makes carbon interests powerful politically. It also explains why decarbonizing is difficult in countries, states or provinces that rely on fossil fuels; compare British Columbia to Alberta, Vermont to West Virginia, or Poland to Denmark, and the role of fossil fuel interests is clear.

The idea of a green economy gained currency after the 2008 financial crisis and the ensuing recession. Governments were eager to stimulate economies with spending, including on clean energy and other green investments. Out of the crisis came an opportunity to promote energy efficiency, renewables, grid upgrades, and other aspects of clean energy. Global investments in clean energy and other environmental goals amounted to nearly 3 trillion dollars over a three- to five-year period, with about 15% aimed at stabilizing or cutting greenhouse gas emissions. In percentage terms, the most spending was in China, South Korea, Germany, France, and the US.[38]

The clean energy case should rest on how it facilitates the achievement of economic and social goals, increases rather than restricts opportunities for durable jobs, and improves well-being – better health, efficient water and materials use, technology change spilling over into other areas, protection of ecosystems and services, social and economic equity, clean air and water, diversified economies, an expanded tax base, and less damage from climate change.[39] Green economy thinking is clear in the European Green Deal.[40] Its goals are to eliminate pollution and integrate economic, environmental, and social decision-making. It aims to raise living standards, guarantee jobs, promote economic and social equity, empower communities, support climate justice, and limit the average temperature rise to 1.5 degrees.

Policymakers viewed the 2020 Covid-19 pandemic as another chance for a bite at the green spending apple. This was captured in President Biden's plans to rebuild by investing in clean energy.[41] An analysis of spending by the largest emitters early in 2021 found that "only the EU has committed a meaningful share of its stimulus" to clean energy; the US, China, and India had allocated a "negligible" 1% or less.[42] This study defined clean energy spending as "any measure that supports energy efficiency, zero-emission energy gener-ation or equipment (e.g., renewable energy investments, subsidies for electric or zero-emission vehicles) as well as infrastructure necessary for reaching long-term net-zero targets (e.g., transit and rail invest-ments, EV charging infrastructure, and forest restoration)."[43]

In the wake of the 2008 financial crisis, the top clean energy spenders by amount were the US, South Korea, and Japan. This time, they were the EU, Germany, and France. Analyses of green stimulus spending after the 2009 crisis find that these investments yielded returns for clean energy. China's PV capacity expanded by a factor of twenty between 2007 and 2011, while US wind turbine production grew 72% from 2007 to 2012. In the EU, each $1.00

invested boosted GDP by \$1.50. Green spending programs generated some 900,000 job-years in the US and 156,000 in South Korea.[44] Of course, it should not take an economic crisis to provoke green investments.

Clean energy proponents are searching for a political strategy. Although the green economy concept is generally accepted as good politics, linking energy and economic goals in positive ways, there is divergence over how to frame the issues. In the US in 2019, several progressive members of Congress proposed linking climate and energy action with ambitious social reform. The Green New Deal called for investment in energy, green infrastructure, mass transit, and other programs.[45] It had broad goals: universal health care; guaranteed jobs; affordable, safe housing; higher education open to all. As one writer put it, this linking of decarbonization with sweeping social change was a way of "mobilizing collective action through a utopian vision of the progressive agenda."[46]

Overcoming carbon lock-in, in its political as well as physical and social aspects, means that something has to punctuate the policy equilibrium. Recall from Chapter 1 that the punctuated equilibrium framework developed as an alternative to incrementalism for explaining disruptive policy change. It applies to the clean energy transition, which requires disruption in the old equilibrium based on fossil fuel dominance. The external forcing event of global climate change, augmented by the health and ecological damages of carbon-based energy, have to disrupt the policy status quo.

This is already occurring. Indeed, the engagement of the international community in climate mitigation and active clean energy agendas in many countries and regions is a sign of disruption. That the carbon-based energy system and the policies supporting it are being contested is not in doubt. The political subsystems on which the politics and policies of carbon lock-in rest are being disrupted. The transition becomes more difficult as policies become less distributive and more regulatory, causing more political conflict. Energy policies in many countries and international forums are designed less to maintain the existing energy system than to accelerate the transition.

The political challenge is to build a durable coalition to displace carbon lock-in. It is not an easy path, and there will be stops and starts. In their book on renewable energy politics, Michael Aklin and Johannes Urpelainen warn that early success with renewables mobilizes the opposition. Ultimately, success will depend on public opinion, government ideology, and industry clout.[47] Clean energy interests must "either disarm or appease the powerful opponents of renewable energy policies."[48] Whatever the strategy, politics will determine the success or failure of the transition.

Keeping Options on the Table

The legendary baseball player Yogi Berra once said: "It's tough to make predictions, especially about the future." In the 1950s, nuclear power seemed to be the key to energy's future, a technology that was "too cheap to meter." In 2009, it was hard to envision that PV costs would fall by 90% and it would be the fastest-growing source of electricity. Making predictions is indeed notoriously difficult.

We cope with uncertainty by hedging our bets, by adopting strategies that will protect us if even reasonable expectations are not fulfilled. Insurance is one way of doing this. People take out life, health, or accident insurance in case things do not go as expected. Hedging bets is especially necessary when there is little opportunity for do-overs – for going back and trying another strategy if the first does not pan out. With the energy transition, there is a limited opportunity for do-overs. The world has to get it as close to right as possible and find ways to hedge its bets. One way of doing this is by keeping all reasonable options on the table.

Managing this transition is in large part a matter of managing the risks of uncertainty. Nuclear technology is an obvious case in point, although CCS and BECCS illustrate it as well. Many clean energy advocates argue that the world can reliably run on wind, solar, and water sources later in this century. But what if it cannot? What if the difficulty of integrating renewables into grids becomes insurmountable? What if opposition to wind and solar siting delays their scale-up? What if silicon-based PV technology becomes so locked in that even more efficient options are not adopted at scale? What if offshore wind platforms do not perform well enough and investors back away? What if battery storage, vehicle-to-grid technologies, or smart grids do not progress as rapidly as expected?

These are only a few of the uncertainties that may push a narrowly conceived transition off course. Having to go back at mid-century and revive options taken off the table now would be harmful – to the vulnerable groups most harmed by climate change, to coastal cities like Miami and Manila inundated by rising sea levels, to billions of people breathing air polluted by fossil fuels, and to indigenous communities whose ecological capital is destroyed by oil extraction. This does not mean that any technology or source should be immune from criticism. Even wind and PV, the workhorses, have limitations. Fair-minded people have doubts about many aspects of clean energy. The safety, reliability, and costs of all options should be considered. At the same time, eliminating promising technologies is risky. As we have seen, despite the skepticism about nuclear power, there are prospects for advanced designs using small, modular reactors with passive safety. There are reasons to doubt CCS and BECCS, yet neither has been fully tested in markets, and many scenarios see carbon removal as critical for decarbonizing and achieving carbon neutrality.

The clean energy transition will not be smooth; nor will it be seamless. If it is perceived as leading to shortages or politically unacceptable price hikes, it could falter. In the fall of 2021, as this chapter was being revised, *The Economist* warned of "the first big energy shock of the green era." Since May that year, the price of oil, coal, and gas had risen by 95%, and the UK had turned some shuttered coal plants back on. Blackouts were hitting China and India. Russia was using Europe's natural gas dependence as a source of political leverage. Disruptions occurred due to insufficient wind in Europe, droughts affecting hydro in Latin America, and floods blocking coal deliveries in Asia. As democracies cut back on fossil fuels, demand shifted "to autocracies with fewer scruples and lower costs."[49] Global investment was half of that needed to meet net-zero targets.

This is not a warning against a transition. It is underway, and it should be accelerated. It is a caution on how it must be strategic, well coordinated, and use a range of options. Governments should move it along, connect the pieces, provide funding, create incentives, and facilitate global cooperation. Making this happen is one of the most pressing challenges of this century.

Pessimism or Optimism?

People often ask whether I lean toward optimism or pessimism on the clean energy transition. On the positive side, there is the rapid growth of wind and solar, progress on green hydrogen and storage, growing mobilization around the world on climate action, and growth in clean energy products. On the negative, there is fossil fuel dominance, the technology challenges still to be met, and the growing global energy appetite. There is ample evidence for both optimists and pessimists.

An optimistic view

One can imagine a bright future in a carbon-neutral world. In the best clean energy future, global investment grows to the levels urged by the IEA and IRENA. An era of global stability delivers sound governance that recognizes clean energy's benefits. Public investments and policy support swift innovation and scale-up; technologies for battery storage, green hydrogen, advanced nuclear, enhanced geothermal, innovative PV materials, and cost-effective carbon removal grow rapidly. Barriers to offshore wind, technology-smart grids, and green hydrogen economies fall away.

At a policy level, governments learn to combine mandates with financial incentives and market creation to improve energy productivity, make low-carbon sources the dominant means of generating electricity, electrify nearly all of transport and industry, and meet remaining needs with low- or zero-carbon technology and practices. Governments rethink land use and mobility. The Paris goals are met, and countries reach new levels of cooperation to avoid the worst climate impacts. Electricity access in low-income countries grows, enabled by distributed energy, support from rich countries, and sound governance. Risks to health and climate fall rapidly.

Within countries, the political and economic influence of renewable energy reaches a tipping point. Many clean energy advocates fear the prospect of technology lock-in, but at least the politics of energy are transformed. A door opens on a new era of investment and innovation, domestic political action, and global cooperation. This occurs in ways that are fair to all – to low-income households struggling to afford energy, to communities affected by climate change and pollution, to workers who suffer in the transition. Investments and policy designs account for equity and fairness. The transition is just. The case for a green economy made by clean energy advocates is validated: it delivers a better life and world, with social, economic, and health benefits that are beyond dispute.

A pessimistic view

In contrast, things may not go well. Consider a future in which extreme parties gain control of governments, or external events (a global depression, a pandemic, regional wars) lead to economic and political instability, undermine global cooperation, and distract governments and the public. It is difficult for governments and the private sector to make investments and change policies in such circumstances. One risk is the rise of right-wing populist parties in many countries. They tend to favor fossil fuels over renewables and disparage global cooperation on issues like climate change.[50]

Whether as a result of economic instability, regional conflicts, or regressive politics, we could face a future in which the clean energy transition stalls or is reversed. In this case, governments fail to exert pressure for decarbonizing. Although a handful of countries price carbon and push wind and solar, most perpetuate carbon lock-in. Energy demand rises, despite efficiency's benefits, and investments in research and infrastructure fall well short. Renewables merely keep pace with rising energy use; the use of fossil fuels in transport and industry fall only slightly. Although there are marginal gains in carbon intensity, emissions grow steadily. Global cooperation all but collapses; the Paris goals are an historical curiosity. Short-term growth drives political agendas, and carbon lock-in persists.

Of course, clean energy is not the only casualty. The factors limiting the energy transition affect other issues. The collective action on which planetary and human well-being depend falls away – action on water sustainability, ecosystem protection, public health, and more. The world is set on a course in which the effects of unguided growth at some point will exceed planetary capacities. Building a clean energy system is one of many sustainability challenges, but it is a principal one.

These are best and perhaps worst scenarios. From a clean energy perspective, the optimistic version is a happy one, and the bad one could even be worse. The challenge for current and future generations of citizens, activists, investors, and policymakers is to realize a world in which the first scenario becomes more likely – to allow the optimists to win. It is hard to predict the future; it is even harder, but more satisfying, to determine that future and make the world a better place.

Guide to Further Reading

Matthew J. Burke and Jennie C. Stephens, "Political Power and Renewable Energy Futures: A Critical Review," *Energy Research and Social Science* 33 (2018), 78–93.

Daniel J. Fiorino, *A Good Life on a Finite Earth: The Political Economy of Green Growth*, Oxford, 2018.

Cornelia Fraune and Michelle Knodt, "Sustainable Energy Transformations in the Age of Populism, Post-Truth Politics, and Local Resistance," *Energy Research and Social Science* 43 (2018), 1–7.

Robert Fri, "From Energy Wish Lists to Technological Realities," *Issues in Science and Technology* 23 (2006), 63–8.

Glossary

Glossary words are emboldened on first usage in the text.

Advanced nuclear reactors: These typically are small reactors that rely on passive safety systems, are modular in design to reduce costs, generate little or no waste, and offer more flexibility in use than do existing technologies.

Capacity factor: The ratio of actual to potential electricity that is available from a technology or energy source. A higher capacity factor means that the source may be counted on to produce useful electricity more of the time. Fossil fuels and nuclear sources generally have higher capacity factors than renewables like wind and solar.

Carbon capture and storage: A set of technologies for removing carbon from waste streams and storing it underground, in oceans, or elsewhere for indefinite periods of time. When used to manufacture useful products it is termed carbon capture, utilization, and storage (CCUS). When linked to biomass that creates a carbon sink it is termed bioenergy with carbon capture and storage (BECCS).

Carbon intensity: The carbon dioxide emissions associated with producing a given unit of economic output in a society, geographic area, or economic sector. The higher the carbon intensity, the more the emissions released to produce a unit of output.

Carbon lock-in: This describes the dependence of societies and economies around the world on carbon-based energy sources as a consequence of some two centuries of accumulated choices. It describes the technologies, physical infrastructure, political power, public policies, and consumer expectations associated with the existing, fossil-fuel-based energy system.

Carbon neutrality: The state in which human activity is not adding additional carbon dioxide to the atmosphere, due to dramatic reductions in emissions and the expansion of technological and natural means of carbon capture and removal. The goal is to balance the books with net-zero carbon.

Carbon removal: The practice of removing carbon dioxide directly from the atmosphere through natural methods like afforestation and soil carbon management, or through technologies such as direct air capture.

Economies of scale: The more units of a product (such as chemical storage batteries or battery electric vehicles) that are manufactured, the lower the costs of additional units are likely to be. This is part of what makes the scale-up of clean energy technologies possible.

Electrification: One of the pillars of decarbonizing the global energy system, electrification is the practice of converting end uses of fossils fuels in transport, industry, and buildings to electricity generated by renewable energy.

Energy democracy: For energy democracy advocates, the clean energy transition is not just a technical process of change but a social one. Energy systems are embedded in larger social, political, and economic systems. In this view, a clean energy transition is an opportunity to redistribute political power and create more empowered, equitable societies.

Energy intensity: The amount of energy required to produce a unit of economic output in a society, geographic region, or economic sector. As technologies and practices improve, energy intensity tends to decline, which means that energy productivity tends to increase.

Grid integration: As the use of renewable sources for generating electricity increases, they have to be integrated into systems for meeting electricity needs. The variable characteristics of sources like wind and solar pose distinctive challenges that are captured in the term grid integration.

Just transition: This describes the social and economic fairness of the transition to a clean energy system. As with any process of social and technological change, this transition causes dislocation and disruptions (such as lost fossil fuel industry jobs) that should be accounted for.

Levelized cost of electricity: A tool for comparing the unit costs of generating electricity from different sources. It describes the total life-cycle market costs of producing a given unit of electricity, such as a kilowatt-hour. It generally does not account for the social costs of energy from different sources.

Rebound effect: When people and organizations use less energy, it becomes economically feasible to use more of it. This term describes the phenomenon in which energy efficiency technologies enable more energy use, offsetting to some degree the benefits gained through efficiency and conservation.

Renewables: Energy sources that rely on natural sources and are not exhausted by use. They generally include solar, wind, ocean, geothermal, and many biomass sources. They cause considerably less pollution, health damage, and ecosystem disruption than do fossil fuels.

Smart grids: Grid systems that incorporate advanced information and communications technologies to promote efficiency, integration of variable renewables, grid flexibility, collaborative grids, and better management of energy demand. Analogous to a smart phone or television.

Social cost of carbon: This term describes the social cost of emitting one ton of carbon dioxide, usually expressed in US dollars. It includes such effects as extreme weather, sea-level rise, droughts, health damages, and losses in ecosystems and biodiversity.

Social costs: The monetized costs of damages that are not accounted for in market transactions, such as the health effects of air pollution, damages to ecosystems, and the costs of global climate change.

Value deflation: This occurs when so much wind and solar energy is available at certain times that their marginal prices in electricity markets fall to or near to zero, undermining their economic viability.

Variable sources: Renewables like wind and solar are variable in that they are not always available when needed. They are *flows* of energy. Sources like coal and natural gas provide *stocks* of energy that are seen as firm – on hand and available when needed. Energy storage is thus critical for variable sources.

Notes

1 The Energy Landscape

1 European Commission, *A European Green Deal*. Accessed September 7, 2021, https://ec.europa.eu/info/strategy/priorities-2019–2024/european-green-deal_en
2 Steven Lee Myers, "China's Pledge to Be Carbon Neutral by 2060: What It Means." *New York Times*, September 23, 2020.
3 Smriti Mallapaty, "How China Could Be Carbon Neutral by Mid-Century." *Nature*, October 22, 2020.
4 Clea Schumer, "How National Net-Zero Targets Stack Up after the COP26 Climate Summit." World Resources Institute, November 18, 2021. Accessed December 11, 2021, https://www.wri.org/insights/how-countries-net-zero-targets-stack-up-cop26
5 John Muyskens and Juliet Eilperin, "Biden Calls for 100% Clean Energy by 2035." *Washington Post*, July 30, 2020.
6 Clean Energy States Alliance, "100% Clean Energy Collaborative-Table of 100% Clean Energy States." Accessed June 3, 2021, https://www.cesa.org/projects/100-clean-energy-collaborative/guide/table-of-100-clean-energy-states/
7 DNV GL, *Energy Transition Outlook 2020: Executive Summary*, 21. Accessed June 4, 2021, https://eto.dnv.com/2020
8 Ibid., 17.
9 Ibid., 21.
10 Ibid., 23.
11 IEA, *Net Zero by 2050: A Roadmap for the Global Energy Sector* (May 2021), 14. Accessed June 1, 2021, https://iea.blob.core.windows.net/assets/4482cac7-edd6–4c03-b6a2–8e79792d16d9/NetZeroby2050-ARoadmapfortheGlobalEnergySector.pdf
12 DNV GL, *Energy Transition Outlook 2020*, 10.

13 IRENA, *Global Renewables Outlook: Energy Transformation 2050* (2020), 16. Accessed June 4, 2021, https://www.irena.org/publications/2020/Apr/Global-Renewables-Outlook-2020

14 Ibid., 19.

15 Ibid., 34.

16 REN21, *Renewables 2019: Global Status Report* (2020), 21. Accessed June 4, 2021, https://www.ren21.net/wp-content/uploads/2019/05/gsr_2019_full_report_en.pdf

17 Richard Rhodes, *Energy: A Human History* (Simon & Schuster, 2018).

18 Roger Fouquet, "Historical Energy Transitions: Speed, Prices, and System Transformation." *Energy Research and Social Science* 22 (2016), 12.

19 Vaclav Smil, "Moore's Curse and the Great Energy Delusion." *The American*, November/December 2008.

20 Matthew J. Burke and Jennie C. Stephens, "Political Power and Renewable Futures: A Critical Review." *Energy Research and Social Science* 35 (2018), 78–93.

21 Discussed in Anthony Giddens, *The Politics of Climate Change*, 2nd ed. (Polity Press, 2011).

22 University of Calgary, "Energy Education." Accessed June 4, 2021, https://energyeducation.ca/encyclopedia/Electricity

23 World Bank, *Tracking SDG7: The Energy Progress Report, 2020.* Accessed May 31, 2021, https://trackingsdg7.esmap.org/about-us

24 IEA, *Key World Energy Statistics 2021*, September 2021, 6. Accessed December 6, 2021, https://iea.blob.core.windows.net/assets/52f66a88–0b63–4ad2–94a5–29d36e864b82/KeyWorldEnergyStatistics2021.pdf

25 Ibid., 30.

26 Ibid., 8.

27 Ibid., 18.

28 Ibid., 13.

29 EIA, *International Energy Outlook 2021: Case Descriptions.* Accessed December 8, 2021, https://www.eia.gov/outlooks/ieo/pdf/IEO2021_CaseDescriptions.pdf

30 EIA, *International Energy Outlook 2021*, 8. Release Presentation; accessed December 9, 2021, https://www.eia.gov/outlooks/ieo/pdf/IEO2021_ReleasePresentation.pdf

31 Ibid., 14.

32 World Bank, *Tracking SDG7*, 11. A gigajoule is 1 billion joules; a joule is a very small unit of energy.

33 From Hannah Ritchie and Max Roser, *Our World in Data, Energy.* Accessed March 3, 2022, https://ourworldindata.org/energy

34 From the WRI's *Climate Watch.* Accessed December 11, 2021, https://www.climatewatchdata.org/ghg-emissions?end_year=2018&start_year=1990

35 Ibid.

36 *Our World in Data*, "Who Has Contributed Most to CO2 Emissions?"

Accessed June 15, 2021, https://ourworldindata.org/contributed-most-global-co2

37 Jonathan Harris and Brian Roach, *Environmental and Natural Resource Economics: A Contemporary Approach*, 4th ed. Tufts University: Global Development and Environment Institute, 2016, Chapter 11, "Energy: The Great Transition."

38 IEA, *World Energy Outlook 2021*, 15. Accessed December 6, 2021, https://iea.blob.core.windows.net/assets/888004cf-1a38–4716–9e0c-3b0e3fdbf609/WorldEnergyOutlook2021.pdf

39 World Bank, "GDP Growth." Accessed October 5, 2021, https://data.worldbank.org/indicator/NY.GDP.MKTP.KD.ZG

40 Robin Cowan, "Nuclear Power Reactors: A Study in Technological Lock-In." *Journal of Economic History* 50 (1990), 541–67.

41 Max Richter, "Which Countries Achieved Economic Growth, and Why Does It Matter?" *Our World in Data*, June 25, 2019, https://ourworldindata.org/economic-growth-since-1950

42 Jessica Lambert et al., "Energy, EROI, and Quality of Life." *Energy Policy* 64 (2014), 153.

43 Ibid., 158.

44 Gregory C. Unruh, "Understanding Carbon Lock-In." *Energy Policy* 28 (2000), 818.

45 Gretchen Bakke, *The Grid: The Fraying Wires Between Americans and Our Energy Future* (Bloomsbury, 2016), xii.

46 Total cumulative federal oil and gas subsidies from 1918 to 2009 amounted to $447 billion. Nancy Pfund and Ben Healey, "What Would Jefferson Do? The Historical Role of Federal Subsidies in Shaping America's Energy Future." *DBL Investors*, September 2011. Accessed May 31, 2021, https://www.dbl.vc/wp-content/uploads/2012/09/What-Would-Jefferson-Do-2.4.pdf

47 Laura Hale, "Happy 60th Birthday, Interstate Highway System." *2021 Report Card for America's Infrastructure*, June 29, 2016. Accessed May 31, 2021, https://infrastructurereportcard.org/happy-60th-birthday-interstate-highway-system/

48 Kirsten Jenkins, "Setting Energy Justice Apart from the Crowd: Lessons from Environmental and Climate Justice." *Energy Research and Social Science* 39 (2018), 117–21.

49 An example is Daniela Stevens, "The Influence of the Fossil Fuel and Emission-Intensive Industries on the Stringency of Mitigation Policies: Evidence from the OECD Countries and Brazil, Russia, India, Indonesia, China, and South Africa," *Environmental Policy and Governance* 29 (2019), 279–92.

50 Unruh, "Understanding Carbon Lock-In." Also by Unruh, "Escaping Carbon Lock-In." *Energy Policy* 30 (2002), 317–25.

51 From Stephen Peake, *Renewable Energy: Power for a Sustainable Future*, 4th ed. (Oxford University Press, 2018) and Lazard's

Levelized Cost of Energy Analysis, versions 12.0 and 13.0 (2018 and 2019).

52 For details, see Richard Heinberg and David Fridley, *Our Renewable Future: Laying the Path for One Hundred Percent Clean Energy* (Island Press, 2016), 117–19.

53 Charles A. S. Hall, Jessica Lambert, and Stephen B. Balogh, "EROI of Different Fuels and the Implications for Society." *Energy Policy* 64 (2014), 150. John W. Day and Charles Hall, *America's Most Sustainable Cities and Regions: Surviving the 21st Century Megatrends* (Springer, 2017), 185–93.

54 IRENA, "Renewables Increasingly Beat Even Cheapest Coal Competitors on Cost." June 2, 2020, https://www.irena.org/newsroom/ pressreleases/2020/Jun/Renewables-Increasingly-Beat-Even-Cheapest-Coal-Competitors-on-Cost

55 Updated annually at Lazard's *Levelized Cost of Electricity and Levelized Costs of Storage*. A recent one is at https://www.lazard.com/perspective/ levelized-cost-of-energy-and-levelized-cost-of-storage-2020/

56 K. S. Gallagher et al., "The Energy Technology Innovation System." *Annual Review of Environment and Resources* 37 (2012), 137–62.

57 Warren Cornwall, "Renewable Power Surges as the Pandemic Scrambles Global Energy Outlook, New Report Finds." *Science*, April 30, 2020.

58 Susanna Twidale, "Green Energy Ratchets Up Power During Coronavirus Epidemic." *Reuters*, July 22, 2020.

59 Santosh Raikar and Seabron Adamson, *Renewable Energy Finance: Theory and Practice* (Elsevier, 2020), 214. The description in this box draws on the definition given in this book.

60 Frank R. Baumgartner and Bryan D. Jones, *Agendas and Instability in American Politics* (University of Chicago Press, 1993). This draws upon M. Masse Jolicoeur, "An Introduction to Punctuated Equilibrium: A Model for Understanding Stability and Dramatic Change in Public Policies." National Center for Collaborative Policy, Montreal, 2018. Accessed August 3, 2021, https://www.ncchpp.ca/docs/2018_ProcessPP_ Intro_PunctuatedEquilibrium_EN.pdf

61 Charles Lindblom, "The Science of Muddling Through." *Public Administration Review* 19 (2) (1959), 79–88.

62 William R. Lowry anticipated this in "Disentangling Energy Policy from Environmental Policy." *Social Science Quarterly* 89 (5) (2008), 1195–211.

63 Corrine Le Quere et al., "Fossil Fuel Emissions in the Post-COVID-19 Era." *Nature Climate Change* 11 (2021), 197.

64 John M. Reilly, Y. H. Henry Chen, and Henry D. Jacoby, "The COVID-19 Effect on the Paris Agreement." *Humanities and Social Sciences Communications*, January 18, 2021, 1.

65 International Energy Agency, "Global CO2 Emissions Rebuild to Their Highest Level in History in 2021," March 8, 2022. Accessed March

31, 2022, https://www.iea.org/news/global-co2-emissions-rebounded-to-their-highest-level-in-history-in-2021

66 Joan Michelson, "Europe Expedites Transition to Clean Energy Due to Ukraine Invasion," *Forbes*, February 26, 2022.

67 Carbon Brief, "Q/A: What Does Russia's Invasion of Ukraine Mean for Energy and Climate Change?" February 25, 2022. Accessed March 20, 2022, https://www.carbonbrief.org/qa-what-does-russias-invasion-of-ukraine-mean-for-energy-and-climate-change

2 Why Clean Energy Matters

1 Roger Fouquet, "Long-Run Trends in Energy-Related External Costs." *Ecological Economics* 20 (2011), 83.

2 Ibid., 84–7.

3 IPCC, *Special Report: Global Warming of 1.5 Degrees C, Summary for Policy Makers*, October 2018. Accessed June 5, 2021, https://www.ipcc.ch/2018/10/08/summary-for-policymakers-of-ipcc-special-report-on-global-warming-of-1–5c-approved-by-governments/

4 Judy Wu, Gaelen Snell, and Hasina Samji, "Climate Anxiety in Young People: A Call to Action." *The Lancet: Planetary Health* 4 (10) (2020), e435–e436.

5 WHO, "How Air Pollution is Destroying Our Health." Accessed March 3, 2022, https://www.who.int/news-room/spotlight/how-air-pollution-is-destroying-our-health

6 Anjum Hajat, Charlene Hsia, and Marie S. O'Neill, "Socioeconomic Disparities in Air Pollution: A Global Review." *Current Environmental Health Reports* 2 (2015), 440–50.

7 Examples are IRENA, *Renewable Energy Benefits: Measuring the Economics* (2016) and, for the US, Devashree Saha and Joel Jager, *America's New Climate Economy: A Comprehensive Guide to the Economic Benefits of Climate Policy in the United States*, World Resources Institute, February 2020.

8 Michael Greenstone and Adam Looney, "Paying Too Much for Energy? The True Costs of Our Energy Choices." *Daedalus* 141 (2) (2012), 10–30; Erica Gies, "The Real Cost of Energy." *Nature*, November 29, 2017.

9 A summary is given by the Union of Concerned Scientists (UCS), *The Hidden Costs of Fossil Fuels*, August 30, 2016. Accessed August 5, 2021, https://www.ucsusa.org/resources/hidden-costs-fossil-fuels.

10 Ibid. This is based on the UCS summary.

11 Richard Schiffman, "A Troubling Look at the Human Toll of Mountaintop Removal Mining." *Yale Environment 360*, November 21, 2017.

12 Daniel Raimi, *The Fracking Debate: The Risks, Benefits, and Uncertainties of the Shale Revolution* (Columbia University Press, 2017).

13 John P. Rafferty, "9 of the Biggest Oil Spills in History." *Britannica*.

Accessed December 15, 2020, https://www.britannica.com/list/9-of-the-biggest-oil-spills-in-history

14 Akshat Rathi, "You Probably Have No Idea How Much Water Is Needed to Produce Electricity." *Quartz*, August 8, 2018.

15 Earth Justice, "The Coal Ash Problem." Accessed July 14, 2020, https://earthjustice.org/features/the-coal-ash-problem

16 Paul R. Epstein et al., "Full Cost Accounting for the Life Cycle of Coal." *Annals of the New York Academy of Sciences* 1219 (2011), 73–98.

17 National Academy of Sciences, *Hidden Costs of Energy: Unpriced Consequences of Energy Production and Use* (National Academies Press, 2010).

18 Ibid., 6.

19 Ibid., 21.

20 IPCC, *Climate Change 2014: Impacts, Adaptation, and Vulnerability: Summary for Policy Makers* (5th Assessment), 5.

21 From the US Global Change Research Program's *Climate Science Special Report*, November 2017.

22 IPCC Press Release, *Climate Change is Widespread, Rapid, and Intensifying*, August 9, 2021. Accessed December 14, 2021, https://www.ipcc.ch/2021/08/09/ar6-wg1-20210809-pr/

23 US Environmental Protection Agency, "Climate Change Impacts." Accessed September 7, 2021, https://19january2017snapshot.epa.gov/climate-impacts_.html

24 US Environmental Protection Agency, "Global Greenhouse Gas Emission Data." Accessed June 7, 2021, https://www.epa.gov/ghgemissions/global-greenhouse-gas-emissions-data

25 Will Steffen, "A Truly Complex and Diabolical Policy Problem." In John S. Dryzek, Richard P. Norgaard, and David Schlosberg (eds), *Oxford Handbook of Climate Change and Society* (Oxford University Press, 2011); see also Daniel J. Fiorino, *Can Democracy Handle Climate Change?* (Polity Press, 2018).

26 For a review, see Brad Plumer and Nadia Popovich, "Yes, There Has Been Progress on Climate. No, It's Not Nearly Enough." *New York Times*, October 25, 2021.

27 Global Carbon Project, "Carbon Budget 2021." November 4, 2021, update. Accessed December 16, 2021, https://www.globalcarbonproject.org/carbonbudget/index.htm

28 United Nations Environment Program, *Emissions Gap Report 2020*, xxi.

29 Quoted in Alan Neuhauser, "Protecting the Earth for Public Health." *U.S. News & World Report*, July 6, 2018. Accessed April 10, 2020, https://www.usnews.com/news/the-report/articles/2018-07-06/gina-mccarthy-on-how-the-environment-is-a-public-health-issue

30 Michael Greenstone, Testimony Before the House of Representatives Committee on Oversight and Reform, "The Devastating Impacts

of Climate Change." August 8, 2020. Accessed March 3, 2022, https://epic.uchicago.edu/wp-content/uploads/2020/08/Greenstone_Testimony_08052020.pdf?mc_cid=5fca2469fa&mc_eid=af3b33de76

31 Peter Howard, Iliana Paul, and Jason Schwartz, *The Social Cost of Carbon and State Policy*, Institute for Policy Integrity, 2017. Accessed March 3, 2022, https://policyintegrity.org/files/publications/SCC_State_Guidance.pdf

32 Greenstone, Testimony Before the House of Representatives, 6.

33 Jeff Goodell, "Can We Survive Extreme Heat?" *Rolling Stone*, August 27, 2019.

34 NOAA, "Climate Change: Sea Level" (October 2021). Accessed December 14, 2021, https://www.climate.gov/news-features/understanding-climate/climate-change-global-sea-level

35 Timothy M. Lenton et al., "Climate Tipping Points – Too Risky to Bet Against." *Nature*, November 27, 2019.

36 WHO, *Air Pollution*. Accessed August 10, 2021, https://www.who.int/health-topics/air-pollution#tab=tab_1

37 *Science Daily*, March 5, 2020. The study is Thomas Munzel et al., "Loss of Life Expectancy from Air Pollution Compared to Other Risk Factors: A Worldwide Perspective." *Cardiovascular Research* 2020. Accessed December 14, 2021. See https://www.sciencedaily.com/releases/2020/03/200305135048.htm

38 Both quotes are from p. 7 of UNICEF, *Silent Suffocation in Africa: Air Pollution Is a Growing Menace, Affecting the Poorest Children the Most* (2015).

39 Ioannis Manisalidis et al., "Environmental Impacts of Air Pollution: A Review." *Frontiers in Public Health* 8 (2020), 1–13.

40 Michael Greenstone and Claire Fan, *Air Quality Annual Index*, Annual Update, July 2020. Accessed March 3, 2022, https://aqli.epic.uchicago.edu/wp-content/uploads/2020/07/AQLI_2020_Report_FinalGlobal-1.pdf

41 On India, see Steven Bernard and Amy Kazmin, "Dirty Air: How India Became the Most Polluted Country on Earth." *Financial Times*, December 11, 2018.

42 J. Lelieveld et al., "Effects of Fossil Fuel and Total Anthropogenic Emission Removal on Public Health and Climate." *PNAS* 116 (2019), 7192–7.

43 Jonathan J. Buonocore et al., "Climate and Health Benefits of Increasing Renewable Energy Deployment in the United States." *Environmental Research Letters* 14 (2019).

44 Drew Shindall, Greg Faluvegi, Karl Seltzer, and Cary Shindall, "Quantified, Localized Health Benefits of Accelerated Carbon Dioxide Emissions Reductions." *Nature Climate Change* 8 (2018), 291–5.

45 Lauri Myllyvirta, *Quantifying the Economic Costs of Air Pollution from Fossil Fuels*, Centre for Research on Energy and Clean Air (2020).

Accessed July 10, 2021, https://energyandcleanair.org/wp/wp-content/uploads/2020/02/Cost-of-fossil-fuels-briefing.pdf

46 Tara Failey, "Poor Communities Exposed to Elevated Air Pollution Levels." *Global Health Environmental Newsletter*, National Institute of Environmental Health Sciences, April 2016. The global review is Anjum Hajat, Charlene Hsia, and Marie S. O'Neil, "Socioeconomic Disparities and Air Pollution: A Global Review." *Current Environmental Health Reports* 4 (December 2015), 440–50.

47 American Lung Association, "Disparities in the Impact of Air Pollution." Accessed June 8, 2021, https://www.lung.org/clean-air/outdoors/who-is-at-risk/disparities

48 Beth Gardiner, *Choked: Life and Breath in the Age of Air Pollution* (University of Chicago Press, 2019), 88.

49 Ibid., 125.

50 Ibid., 63.

51 Ibid., 258.

52 Peter Howard and Derek Sylvan, *Expert Consensus on the Economics of Climate Change*, Institute for Policy Integrity, New York University School of Law, 2015.

53 Kevin Rennert and Cora Kingdon, *Social Cost of Carbon 101*, Resources for the Future, August 2019. Accessed November 21, 2020, https://media.rff.org/documents/SCC_Explainer.pdf

54 Interagency Working Group on the Social Cost of Carbon, *Social Cost of Carbon for Regulatory Impact Analysis* (2010). Accessed November 20, 2020, https://www.epa.gov/sites/production/files/2016–12/documents/scc_tsd_2010.pdf

55 Brad Plumer, "Trump Put a Low Cost on Carbon Emissions. Here's Why It Matters." *New York Times*, August 23, 2018.

56 Katharine Ricke, Laurent Drouet, Ken Caldeira, and Massimo Tavoni, "Country-Level Social Cost of Carbon." *Nature Climate Change* 8 (10) (2018), 895–900.

57 Sebastian Helgenberger and Martin Janicke, *Mobilizing the Co-Benefits of Climate Mitigation: Connecting Opportunities with Interests in the New Energy World of Renewables*, Institute for Advanced Sustainability Studies, Potsdam, July 2017.

58 Victoria Mausten, Michael Daly, and Liam Delaney, "The Scarring Effects of Unemployment on Psychological Well-Being Across Europe." *Social Science Research* 72 (May 2018), 146–69.

59 Adam Mayer, "A Just Transition for Coal Miners? Community Identity and Support from Local Policy Actors." *Environmental Innovation and Societal Transformations* 28 (2018), 1–13.

60 Van Jones, *The Green-Collar Economy: How One Solution Can Fix Our Two Biggest Problems* (Harper One, 2008), 14 and 12.

61 The studies are Robert Pollin, James Heintz, and Heidi Garrett-Peltier, *The Economic Benefits of Investing in Clean Energy* (Washington, DC:

Center for American Progress, 2009), and, by the same authors plus Bracken Hendricks, *Green Growth: A US Program for Controlling Climate Change and Expanding Job Opportunities* (Washington, DC: Center for American Progress, 2014).

62 OECD, *Towards Green Growth* (2011), 91 and 92. Accessed April 3, 2022, https://read.oecd-ilibrary.org/environment/towards-green-growth_9789264111318-en#page111

63 Heidi Garrett-Peltier, "Green versus Brown: Comparing the Employment Impacts of Energy Efficiency, Renewable Energy, and Fossil Fuels Using an Input-Output Model." *Economic Modelling* 61 (2017), 439.

64 IRENA, *Renewable Energy and Jobs: Annual Review 2019*. Accessed July 15, 2020, https://www.irena.org/-/media/Files/IRENA/Agency/Publication/2019/Jun/IRENA_RE_Jobs_2019-report.pdf

65 Securing America's Future Energy (SAFE). *The Military Cost of Defending the Global Oil Supply: Issue Brief* (2018). Accessed September 15, 2020, http://secureenergy.org/wp-content/uploads/2020/03/Military-Cost-of-Defending-the-Global-Oil-Supply.-Sep.-18.-2018.pdf

66 On national security co-benefits, see US Department of Energy, *Valuation of Energy Security for the United States*, January 2017. Accessed March 3, 2022, https://www.energy.gov/policy/articles/valuation-energy-security-united-states; and American Council on Renewable Energy, *The Role of Renewable Energy in National Security* (October 2018). Accessed March 3, 2022, https://acore.org/wp-content/uploads/2018/10/ACORE_Issue-Brief_-The-Role-of-Renewable-Energy-in-National-Security.pdf

67 IRENA, *Renewable Energy Benefits: Measuring the Economics*, 9.

68 *REmap* is examined in the next chapter. The job benefits are discussed here.

69 IRENA, *Renewable Energy Benefits: Measuring the Economics*, 24.

70 Ibid., 33.

3 Getting the Carbon Out: Pathways to Decarbonization

1 An example is Samuel Alexander, "Planned Economic Contraction: The Emerging Case for Degrowth." *Environmental Politics* 21 (2012), 349–68.

2 Daniel J. Fiorino, *A Good Life on a Finite Earth: The Political Economy of Green Growth* (Oxford University Press, 2018), 6.

3 Carol Rasmussen, "Emission Reductions from Pandemic Had Unexpected Effects on Atmosphere." NASA News, November 9, 2021, https://climate.nasa.gov/news/3129/emission-reductions-from-pandemic-had-unexpected-effects-on-atmosphere/

4 Bloomberg New Energy Finance, "Emissions and Coal Have Peaked as Covid-19 Saves 2.5 Years of Emissions, Accelerates Energy Transition."

October 27, 2021. Accessed March 3, 2022, https://about.bnef.com/blog/emissions-and-coal-have-peaked-as-covid-19-saves-2-5-years-of-emissions-accelerates-energy-transition

5 Henrik Selin and Stacy D. VanDeveer, "Global Climate Change Governance: Where to Go After Paris?" In Norman Vig and Michael E. Kraft (eds.), *Environmental Policy*, 10th ed. (Washington, DC: CQ Press, 2019), 322–46.

6 Michael Oppenheimer and Annie Petsik, "Article 2 of the UNFCCC: Historical Origins, Recent Interpretations." *Climatic Change* 73 (December 2005), 195–26. The statement is on p. 4 of the UNFCCC at https://unfccc.int/resource/docs/convkp/conveng.pdf. Accessed April 3, 2022.

7 Yun Gao, Xiang Gao, and Xiaohua Zhang, "The 2C Global Temperature Target and the Evolution of the Long-Term Goal of Addressing Climate Change – From the United Nations Framework Convention on Climate Change to the Paris Agreement." *Engineering* 3 (2017), 277.

8 Ibid., 275.

9 Nicholas Stern, *The Economics of Climate Change: The Stern Review* (Cambridge University Press, 2007).

10 IPCC, *Special Report: Global Warming of 1.5C*, Executive Summary of chapter 2. Accessed June 10, 2021, https://www.ipcc.ch/sr15/

11 Global Carbon Project, "Carbon Budget 2021." Accessed June 10, 2021, https://www.globalcarbonproject.org/carbonbudget/

12 A resource on natural carbon sinks is MIT's Natural Climate Solutions Program. See John Fernandez and Marcela Angel, "Protecting and Enhancing Natural Carbon Sinks: Natural Climate and Community Solutions." April 22, 2021. Accessed June 10, 2021, https://solve.mit.edu/articles/protecting-and-enhancing-natural-carbon-sinks-natural-climate-and-community-solutions

13 Ibid.

14 Megan Rowling, "Climate Primer: Why Is Everyone Talking About 'Net Zero'?" Thompson Reuters News, September 21, 2020.

15 Kelly Levin and Chantal Davis, "What Does 'Net-Zero Emissions' Mean? 6 Common Questions Answered." World Resources Institute, September 17, 2019, https://www.wri.org/blog/2019/09/what-does-net-zero-emissions-mean-6-common-questions-answered

16 Clea Schumer, How National Net-Zero Targets Stack Up After the COP26 Climate Summit." WRI. Accessed December 17, 2021, https://www.wri.org/insights/how-countries-net-zero-targets-stack-up-cop26

17 United Nations, "Nationally Determined Contributions." Accessed June 14, 2021, https://unfccc.int/process-and-meetings/the-paris-agreement/nationally-determined-contributions-ndcs/nationally-determined-contributions-ndcs

18 DDDP, *Pathways to Deep Decarbonization*, 2015 Synthesis Report. Sustainable Development Solutions Network and Institute for

Sustainable Development and International Relations, December 2015. Accessed March 3, 2022, https://www.iddri.org/sites/default/files/import/publications/ddpp_2015synthetisreport.pdf

19 Heinberg and Fridley, *Our Renewable Future*, 181–5.
20 Ibid., 184.
21 Ibid.
22 Ibid., 185.
23 Ibid., 196.
24 Mark Z. Jacobson et al. "100% Clean and Renewable Wind, Water, and Sunlight All-Sector Energy Roadmaps for 139 Countries of the World." *Joule* 1 (2017), 108–21.
25 Ibid., 108.
26 Mark Z. Jacobson, et al. "Matching Demand with Supply at Low Cost in 139 Countries Among 20 World Regions with 100% Intermittent Wind, Water, and Sunlight (WWS) for All Purposes," *Renewable Energy* 123 (2018), 236.
27 Ibid., 237.
28 Christopher T. M. Clack et al., "Evaluation of a Proposal for a Reliable Low-Cost Grid Powered with 100% Wind, Water, and Solar." *PNAS* 114 (26), 6722.
29 Bernie Sanders and Mark Jacobson, "The American People – Not Big Oil – Must Decide Our Climate Future." *The Guardian*, April 29, 2017.
30 IRENA, *Global Energy Transformation: A Roadmap* (2019). Accessed March 3, 2022, https://www.irena.org/publications/2019/Apr/Global-energy-transformation-A-roadmap-to-2050-2019Edition
31 Ibid., 4.
32 Ibid., 8.
33 Ibid., 31.
34 Ibid., 33.
35 IEA, *Net Zero by 2050: A Roadmap for the Global Energy Sector*, 13.
36 Ibid., 14.
37 Ibid., 23.
38 Ibid., 16–17.
39 Ibid., 25.
40 Chris Bataille et al., "The Deep Decarbonization Pathways Project: Insights and Emerging Issues." *Climate Policy* 15 (1) (2016), S1–S6.
41 DDPP, *Pathways to Deep Decarbonization, 2015 Synthesis Report*.
42 Christopher Bataille et al. "Net-Zero Deep Decarbonization Pathways in Latin America: Challenges and Opportunities." *Energy Strategy Reviews* 30 (2020), 100510.
43 Ibid., 7.
44 The Zero Carbon Consortium, *America's Zero Carbon Action Plan*, Sustainable Development Solutions Network, 2020.
45 Ibid., 4.
46 Ibid., 30.

47 Ibid.

48 Ibid.

49 This draws on two reports: International Institute for Sustainable Development in Paris (IDDRI), *Policy Lessons on Deep Decarbonization in Large Emerging Economies*, November 2021; Khushboo Goel, *Deep Decarbonization Strategy for India: Balancing Climate Change and Economic Development*, Kleinman Center for Energy Policy, University of Pennsylvania, 2019.

50 Goel, *Deep Decarbonization Strategy for India*, 1.

51 IDDRI, *Policy Lessons on Deep Decarbonization*, 32.

52 Ibid., 30.

53 Ibid., 34.

54 Ibid., 33.

55 Goel, *Deep Decarbonization Strategy for India*, 9.

56 IDDRI, *Policy Lessons on Deep Decarbonization*, 36.

57 Gayathri Vaidyanathan, "Scientists Cheer India's Ambitious Zero-Carbon Climate Change Pledge." *Nature*, November 5, 2021.

58 IDDRI, *Policy Lessons on Deep Decarbonization*, 27.

59 For a discussion, see K. S. Gallagher et al., "The Energy Technology Innovation System." *Annual Review of Environment and Resources* 37 (2012), 137–62.

4 The Invisible Resource: Energy Efficiency

1 That is the forecast of the US Energy Information Administration: "EIA Projects Nearly 50% Increase in World Energy Usage by 2050, Led by Growth in Asia." Based on EIA's *International Energy Outlook 2019*. Accessed June 19, 2021, https://www.eia.gov/todayinenergy/detail.php?id=41433

2 IEA, *Energy Efficiency 2018: Analysis and Outlook to 2040* (2018), 13. Accessed March 3, 2022, https://www.iea.org/reports/energy-efficiency-2018

3 Steven Nadel and Lowell Ungar, *Halfway There: Energy Efficiency Can Cut Energy Use and Greenhouse Gas Emissions in Half by 2050* (2019). Accessed June 22, 2021, https://www.aceee.org/sites/default/files/publications/researchreports/u1907.pdf

4 The Zero Carbon Consortium, *America's Zero Carbon Action Plan*, Sustainable Development Solutions Network, 2020, 16.

5 ACEEE, *The International Energy Efficiency Scorecard* (2018), v. Accessed March 3, 2022, https://www.aceee.org/portal/national-policy/international-scorecard

6 EIA, "Use of Energy Explained." Accessed March 3, 2022, https://www.eia.gov/energyexplained/use-of-energy/efficiency-and-conservation.php

7 IEA, *Energy Efficiency 2018*, 22.

8 EIA, "Global Energy Intensity Continues to Decline." Accessed March 3, 2022, https://www.eia.gov/todayinenergy/detail.php?id=27032

9 World Bank, *Energy Efficiency in Russia: Untapped Reserves* (2014), see p. 27. This report notes that "Russia's current energy inefficiency is equal to the annual primary energy consumption of France" (p. 5).

10 A useful visual is *Our World in Data*, "Global Direct Primary Energy Consumption." Accessed June 18, 2021, https://ourworldindata.org/grapher/global-primary-energy?country=~OWID_WRL

11 IEA, *Energy Efficiency 2018: Analysis and Outlook to 2040* (2018), 23.

12 Marilyn Brown and Yu Wang, *Green Savings: How Policies and Markets Drive Energy Efficiency* (Praeger, 2015), 8.

13 Todd D. Gerarden, Richard D. Newell, and Robert N. Stavins, "Assessing the Energy Efficiency Gap." National Bureau of Economic Research, 2015. Accessed July 20, 2021, https://www.nber.org/system/files/working_papers/w20904/w20904.pdf

14 Jesse Jenkins, "Are Rebound Effects a Problem for Energy Efficiency?" The Energy Collective Group, October 16, 2014. Accessed March 3, 2022, https://energycentral.com/c/ec/are-rebound-effects-problem-energy-efficiency

15 Ibid.

16 EIA, *International Energy Outlook 2021: Release Presentation.*

17 EIA, *International Energy Outlook 2019*, 17.

18 Ibid., 28.

19 Ibid., 52.

20 Ibid., 68.

21 IEA, *Energy Efficiency 2018.*

22 "Energy Subsidies: A Costly Mistake." *The Economist*, May 19, 2015.

23 The ACEEE factsheet is at https://www.aceee.org/sites/default/files/halfway-there-0919.pdf. Accessed March 3, 2022.

24 Herman E. Daly, *Steady-State Economics*, 2d ed. (Island Press, 1991).

25 On degrowth, see Samuel Alexander, "Planned Economic Contraction: The Emerging Case for Degrowth." *Environmental Politics* 21 (3) (2012), 349–68.

26 Richard Wilkinson and Kate Pickett, *The Spirit Level: Why Equality Is Better for Everyone* (Penguin, 2009).

27 Carol Graham, *The Pursuit of Happiness: An Economy of Well-Being* (Brookings Institution, 2011).

28 A version of this is at Jeffrey Pfeffer, "'Efficient Market' Thinking Is Inefficient." *CBS News*. Accessed June 21, 2021, https://www.cbsnews.com/news/efficient-market-thinking-is-inefficient/

29 Brown and Wang, *Green Savings*, 9–10.

30 David B. Goldstein, "Renewables May Be Plunging in Price, But Efficiency Remains the Touchstone in a Clean Energy Transition." *The Electricity Journal* 31 (2018), 16–19.

31 Marilyn Brown, "Innovative Energy-Efficiency Policies: An International

Review." *WIREs Energy and Environment* 4 (2014), 3. This section draws on this review of efficiency barriers.

32 National Renewable Energy Laboratory (NREL), "Decoupling Policies: Options to Encourage Energy Efficiency Policies for Utilities." Accessed June 21, 2021, https://www.nrel.gov/docs/fy10osti/46606.pdf

33 Richard Thaler and Cass Sunstein, *Nudge: Improving Decisions About Health, Wealth, and Happiness* (Yale University Press, 2008).

34 GBPN, *Energy Efficiency and Energy Savings: A View from the Building Sector*, October 2012. Accessed April 2, 2022, https://tools.gbpn.org/sites/default/files/06.EIU_CaseStudy.pdf

35 International District Heating Association, "What Is Combined Heat and Power?" Accessed November 20, 2021, https://www.districtenergy.org/topics/chp

36 McKinsey, *Pathways to a Low-Carbon Economy* (Version 2 of the Global Greenhouse Gas Abatement Curve, 2009). Accessed June 21, 2021, https://www.mckinsey.com/~/media/mckinsey/dotcom/client_service/sustainability/cost%20curve%20pdfs/pathways_lowcarbon_economy_version2.ashx

37 McKinsey, "A Revolutionary Tool for Cutting Emissions, Ten Years On." April 21, 2017. Accessed March 3, 2022, https://www.mckinsey.com/about-us/new-at-mckinsey-blog/a-revolutionary-tool-for-cutting-emissions-ten-years-on

38 Kenneth Gillingham and James S. Stock, "The Cost of Reducing Greenhouse Gas Emissions." *Journal of Economic Perspectives* 32 (4), 2018, 53–72.

39 Ibid., 65.

40 Brown and Wang, *Green Savings*, discussed 38–49.

41 ACEEE, "How Much Does Energy Efficiency Cost?" Based on studies from 2014 and 2015. Accessed March 3, 2022, https://www.aceee.org/sites/default/files/cost-of-ee.pdf

42 A summary of the many benefits is IEA, *Capturing the Multiple Benefits of Energy Efficiency* (2014). Accessed March 3, 2022, https://www.iea.org/reports/capturing-the-multiple-benefits-of-energy-efficiency

43 Estimate of the Institute for Energy Research, based on EIA and other data, in a "Primer on Energy and the Economy." February 16, 2010. Accessed June 21, 2021, https://www.instituteforenergyresearch.org/uncategorized/a-primer-on-energy-and-the-economy-energys-large-share-of-the-economy-requires-caution-in-determining-policies-that-affect-it/

44 Efficiency's role in helping with renewables integration is stressed by Goldstein, "Renewables May Be Plunging in Price," 18.

45 Ariel Drehobl and Lauren Ross, *Lifting the High Energy Burden in America's Largest Cities: How Energy Efficiency Can Improve Low-Income and Underserved Communities*, ACEEE, 2016, 3–7.

46 See Fiorino, *A Good Life on a Finite Earth*.

47 IEA, *Energy Efficiency 2020*. Accessed March 3, 2022, https://www.iea.org/reports/energy-efficiency-2020

48 ACEEE, *The International Energy Efficiency Scorecard* (2018).

49 This target was tightened in 2018 and is being assessed under the European Green Deal goals. See https://ec.europa.eu/energy/topics/energy-efficiency/targets-directive-and-rules/energy-efficiency-directive_en#content-heading-0. Accessed March 3, 2022.

50 ACEEE, *The International Energy Efficiency Scorecard*, v.

51 Ibid., x.

52 Xavier Labandeira et al., "The Impacts of Energy Efficiency Policies: Meta-Analysis." *Energy Policy* 147 (2020), 10.

53 GBPN, *Energy Efficiency and Energy Savings*, 17.

54 The classic article is Michael E. Porter and Claas van der Linde, "Toward a New Conception of the Environment-Competitiveness Relationship." *Journal of Economic Perspectives* 9 (4) (1995), 97–118.

55 Brian Motherway of the IEA writes: "Energy Efficiency is the first fuel – the fuel you do not have to use – and in terms of supply, it is abundantly available and cheap to extract." IEA, December 19, 2019. Accessed July 10, 2021, https://www.iea.org/commentaries/energy-efficiency-is-the-first-fuel-and-demand-for-it-needs-to-grow

5 Endless Flows: Renewable Energy

1 Oliver Milman, "Renewables Surpass Coal in US Energy Generation for First Time in 130 Years." *The Guardian*, June 3, 2020.

2 Ibid. The quote is from Dennis Wamsted of the Institute for Energy Economics and Financial Analysis.

3 IEA, *Global Energy Outlook*, 1998, 65. Accessed June 20, 2020, https://jancovici.com/wp-content/uploads/2016/04/World_energy_outlook_1998.pdf

4 US Energy Information Administration, *International Energy Outlook 2000*, 93. Accessed April 3, 2022, https://rosap.ntl.bts.gov/view/dot/5104#:~:text=The%20International%20Energy%20Outlook%202000%20%28IEO2000%29%20presents%20an,the%20outlook%20for%20international%20energy%20markets%20through%202020

5 NREL, *Cost Projections for Utility-Scale Battery Storage: 2020 Update* (June 2020). Accessed June 24, 2021, https://www.nrel.gov/docs/fy20osti/75385.pdf

6 BNEF, *New Energy Outlook 2020* (Executive Summary), 16. Accessed June 24, 2021, https://assets.bbhub.io/professional/sites/24/928908_NEO2020-Executive-Summary.pdf

7 BNEF, "Solar, Wind, Batteries to Attract $10 Trillion to 2050, But Cutting Emissions Long-Term Will Require Other Technologies Too." June 18, 2019. Accessed June 24, 2021, https://about.bnef.com/blog/

solar-wind-batteries-attract-10-trillion-2050-curbing-emissions-long-term-will-require-technologies/

8 IRENA, *Renewable Power Generation Costs in 2019*, 13. Accessed April 3, 2022, https://www.irena.org/-/media/Files/IRENA/Agency/Publication/2020/Jun/IRENA_Power_Generation_Costs_2019.pdf

9 Stephen Peake and Bob Everett, "Introducing Renewable Energy." In Peake (ed.), *Renewable Energy*, 2.

10 Up-to-date analyses of costs are in Lazard's *Levelized Costs of Electricity and Storage*. For 2020, see https://www.lazard.com/perspective/levelized-cost-of-energy-and-levelized-cost-of-storage-2020/ Accessed March 4, 2020.

11 Bob Everett et al. "Integrating Renewable Energy." In Peake (ed.), *Renewable Energy*, 569.

12 Godfrey Boyle and Bob Everett, "Solar Photovoltaics." In Peake (ed.), *Renewable Energy*, 115.

13 Discussed in Bob Everett, "Solar Thermal Energy." In Peake (ed.), *Renewable Energy*, 57–114.

14 Current assessments of renewables are at BNEF's *New Energy Outlook* and REN21's *Global Status Report*.

15 This section draws on Derek Taylor, "Wind Energy." In Peake (ed.), *Renewable Energy*, 353–438.

16 The Chokeberry and Sierra Madre Wind Energy Project will have a capacity of 3,000MW. On the growth of wind energy in that state, see Chris Outcult, "Wyoming Confronts Its Wind-Powered Destiny." *Wired*, April 1, 2020. Accessed May 25, 2020, https://www.wired.com/story/wyoming-confronts-wind-powered-destiny/

17 Current data are available from the Global Wind Energy Council, Supply Side Analysis.

18 Sarah Golden, "Has Offshore Wind's Moment Finally Arrived?" *Greenbiz*, March 12, 2021. Accessed June 24, 2021, https://www.greenbiz.com/article/has-offshore-winds-moment-finally-arrived

19 Janet Ramage and Bob Everett, "Hydroelectricity." In Peake (ed.), *Renewable Energy*, 239–89.

20 International Hydropower Association, *The World's Water Battery: Pumped Hydropower Storage and the Clean Energy Transition*, 2018. Accessed June 5, 2020, https://www.hydropower.org/publications/the-world-e2–80–99s-water-battery-pumped-hydropower-storage-and-the-clean-energy-transition

21 James P. Warren, "Deep Geothermal Energy." In Peake (ed.), *Renewable Energy*, 494.

22 Les Duckers and Ned Minus, "Wave Energy." In Peake (ed.), *Renewable Energy*, 439.

23 Jonathan Scurlock et al. "Bioenergy." In Peake (ed.), *Renewable Energy*, 157.

24 On biomass, see Michael S. Hamilton, *Energy Policy Analysis: A*

Conceptual Framework (Armonk: M. E. Sharpe, 2013). He distinguishes *beneficial* from *harmful* biomass.

25 World Bioenergy Association, *Global Bioenergy Statistics 2019*, 16. Accessed March 4, 2022, https://worldbioenergy.org/uploads/191129%20 WBA%20GBS%202019_HQ.pdf

26 Renee Cho, "Is Biomass Really Renewable?" *State of the Planet*, Earth Institute, Columbia University, August 18, 2011. Accessed June 6, 2020, https://blogs.ei.columbia.edu/2011/08/18/is-biomass-really-renewable/

27 Ibid.

28 NREL, *Life Cycle Greenhouse Gas Emissions from Electricity Generation*, January 2013. The fact sheet for this study is at https:// www.nrel.gov/docs/fy13osti/57187.pdf Accessed March 4, 2022.

29 See the chapters in Peake (ed.), *Renewable Energy* (each chapter on a renewable technology includes a brief summary of environmental impacts), and chapter 4 of National Academy of Engineering, *The Power of Renewables: Opportunities and Challenges for China and the United States* (Washington, DC: National Academies Press, 2010). A summary is UCS's *Environmental Impacts of Renewable Energy Technologies*. Accessed March 4, 2022, https://www.ucsusa.org/ resources/environmental-impacts-renewable-energy-technologies

30 Landon Stevens, "The Footprint of Energy." *Strata*, 2017. Accessed July 25, 2020, https://docs.wind-watch.org/US-footprints-Strata-2017.pdf

31 Jesse Jenkins, "How Much Land Does Solar, Wind and Nuclear Energy Require? *Energy Central*, June 25, 2015. Accessed January 10, 2021, https://energycentral.com/c/ec/how-much-land-does-solar-wind-and- nuclear-energy-require

32 Brookings Institution, "Renewables, Land Use, and Local Opposition in the United States." Accessed July 15, 2021, https://www.brookings.edu/ research/renewables-land-use-and-local-opposition-in-the-united-states/

33 EPRI, *Value of Technology in the Electric Power Sector* (August 2018), 1. Accessed March 31, 2022, https://eea.epri.com/pdf/EPRI-P201-Value- of-Technology.pdf

34 Jesse D. Jenkins and Samuel Thernstrom, "Deep Decarbonization of the Electric Power Sector: Insights from Recent Literature." Energy Innovation and Reform Project, 2017, 4. Accessed March 4, 2022, https://www.innovationreform.org/wp-content/uploads/2018/02/ EIRP-Deep-Decarb-Lit-Review-Jenkins-Thernstrom-March-2017.pdf

35 Lucy Brown, "Daily Solar Generation Breaks a Record Set a Year Ago." *Choose*, April 24, 2020. Accessed July 6, 2021, https://www.choose. co.uk/news/2020/solar-energy-breaks-daily-record/

36 Joshua S. Hill, "Solar Has a Record-Breaking Week in Germany, Provides 23% of Generation." *Renew Economy*, April 15, 2020. Accessed July 5, 2021, https://reneweconomy.com.au/solar-has-record-breaking-week- in-germany-provides-23-of-generation-16305/

37 Electric Power Research Institute, "A Primer on Wind and Solar

Value Deflation." Accessed March 4, 2022, https://eea.epri.com/pdf/Back-Pocket-Insights/EPRI-P201-Decreasing-Returns.pdf

38 Varun Sivaram, *Taming the Sun: Innovations to Harness Solar Energy and Power the Planet* (MIT Press, 2018), 57.

39 Ibid., 70.

40 Bethany Frew et al., "Sunny with a Chance of Curtailment: Operating the US Grid with Very High Levels of Solar Photovoltaics." *iScience* 21 (2019), 436–47.

41 Ibid.

42 Amin Al-Habaibeh, "Should We Turn the Sahara Desert Into a Huge Solar Farm?" *The Conversation*, April 30, 2019. Accessed August 21, 2020, https://theconversation.com/should-we-turn-the-sahara-desert-into-a-huge-solar-farm-114450

43 Will de Freitas, "Could the Sahara Turn Africa Into a Solar Superpower?" World Economic Forum, January 17, 2020. Accessed July 10, 2020, https://www.weforum.org/agenda/2020/01/solar-panels-sahara-desert-renewable-energy/

44 Elizabeth Weise and Rick Jervis, "As Climate Threat Looms, Texas Republicans Have a Solution: Giant Wind Farms Everywhere." *USA Today*, October 21, 2019. Accessed June 6, 2020, https://www.usatoday.com/story/news/2019/10/18/texas-wind-energy-so-strong-its-beating-out-coal-power/3865995002/

45 Electricity Transmission Texas. Accessed March 4, 2022, http://www.ettexas.com/Projects/TexasCrez

46 Amy Yee, "Geothermal Energy Grows in Kenya." *New York Times*, February 23, 2018.

47 Gunwoo Kim et al., "An Overview of Ocean Renewable Resources in Korea." *Renewable and Sustainable Energy Reviews* 16 (2012), 2278–88.

48 World Bank, *Regulatory Indicators for a Sustainable World*, 68. Accessed April 3, 2022, https://www.worldbank.org/en/topic/energy/publication/rise---regulatory-indicators-for-sustainable-energy

49 Ibid., 75.

50 Ibid., 78.

51 *BP Statistical Review of Energy 2020*, 54. Accessed April 3, 2022, https://www.bp.com/content/dam/bp/business-sites/en/global/corporate/pdfs/energy-economics/statistical-review/bp-stats-review-2020-full-report.pdf

52 Jan Burck et al., *Climate Change Performance Index: Results, 2021*. Accessed March 4, 2022, https://germanwatch.org/en/19552

53 IEA, "Morocco Renewable Energy Target 2030." Updated October 10, 2019. Accessed September 20, 2020, https://www.iea.org/policies/6557-morocco-renewable-energy-target-2030

54 Jacob Gronholt-Pedersen, "Denmark Sources Record 47% of Power From Wind in 2019." *Reuters*, January 2, 2020. Accessed August 20, 2020, https://www.reuters.com/article/us-climate-change-denmark-

windpower/denmark-sources-record-47-of-power-from-wind-in-2019-idUSKBN1Z10KE
55 This draws on IRENA's *Renewable Energy and Jobs: Annual Review 2019.*
56 Ibid., 12.
57 Janet Ramage and Bob Everett, "Hydroelectricity." In Peake (ed.), *Renewable Energy*, 287.
58 Les Duckers and Ned Minus, "Wave Energy." In Peake (ed.), *Renewable Energy*, 483.
59 "Solar's New Power: Cells Are Getting Better at Converting Sunshine Into Electricity." *The Economist*, May 23, 2020, 68–9. For more detail, see Sivaram, *Taming the Sun*, 150–96.
60 See Giddens, *The Politics of Climate Change*, 77–87.
61 James Murray, "Denmark's Wind Power Vision to Make Its Electricity Sector Fossil-Free by 2030." *NS Energy*, January 20, 2020. Accessed June 10, 2020, https://www.nsenergybusiness.com/features/denmark-electricity-wind-power/
62 Michael Aklin and Johannes Urpelainen, *Renewables: The Politics of a Global Energy Transition* (MIT Press, 2018), 39.
63 Peter Fairley, "Norway Wants to Be Europe's Battery." *IEEE Spectrum*, October 21, 2014. Accessed July 15, 2020, https://spectrum.ieee.org/green-tech/wind/norway-wants-to-be-europes-battery
64 Jason Deign, "Why Norway Can't Become Denmark's Battery Pack." *Greentechmedia*, March 3, 2017. Accessed June 15, 2020, https://www.greentechmedia.com/articles/read/why-norway-cant-become-europes-battery-pack
65 "The Fukushima Disaster Was Not the Turning Point Many Had Hoped." *The Economist*, March 6, 2021.
66 "The Reinvention of Japan's Power Supply Is Not Making Much Headway." *The Economist*, June 13, 2020.
67 IEA, *Japan 2021: Energy Policy Review.* Accessed July 7, 2021, https://www.iea.org/reports/japan-2021
68 This and the following quote are from Sumiko Takeuchi, "Building Toward Large-Scale Use of Renewable Energy in Japan." *The Japan Times*, July 17, 2019.
69 Roger Karapin, "Wind-Power Development in Germany and the United States: Structural Factors, Multiple-Stream Convergence, and Turning Points." In Andreas Duit (ed.), *State and Environment: The Comparative Study of Environmental Governance* (MIT Press, 2014), 111–46.

6 Electrify Everything

1 Ivan Penn, "Poor Planning Left California Short of Electricity in a Heat Wave." *New York Times*, August 20, 2020.

2 California ISO, *Final Root Cause Analysis: Mid-August 2020 Extreme Heat Wave* (2021). Accessed May 15, 2021, http://www.caiso.com/ Documents/Final-Root-Cause-Analysis-Mid-August-2020-Extreme-Heat-Wave.pdf
3 Jesse D. Jenkins, Max Luke, and Samuel Thernstrom, "Getting to Zero Carbon Emissions in the Electric Power Sector." *Joule* 2 (2018), 2506.
4 Bakke, *The Grid*, 26.
5 Jeffrey D. Sachs, *The Age of Sustainable Development* (Columbia University Press, 2015), 82–3.
6 World Bank, "Access to Energy Is At the Heart of Development" (April 2018). Accessed May 20, 2020, https://www.worldbank.org/en/news/ feature/2018/04/18/access-energy-sustainable-development-goal-7
7 David I. Stern, Paul J. Burke, and Stephan B. Bruns, "The Impact of Electricity on Economic Development: A Macroeconomic Perspective." *Oxford Policy Management* (2017), 21. Accessed March 5, 2020, https://escholarship.org/uc/item/7jb0015q
8 Union of Concerned Scientists, "How the Electricity Grid Works." Accessed March 5, 2022, https://www.ucsusa.org/resources/how-electricity-grid-works
9 Ibid.
10 Peter Cramton, "Lessons from the 2021 Texas Electricity Crisis." *Utility Dive*, March 23, 2021. Accessed April 10, 2021, https://www.utilitydive. com/news/lessons-from-the-2021-texas-electricity-crisis/596998/
11 Based on Kathryne Cleary and Karen Palmer, "US Electricity Markets 101." *Resources for the Future*, March 3, 2020. Accessed November 15, 2021, https://media.rff.org/documents/US_Electricity_Markets_101.pdf
12 Lazard's *Levelized Cost of Energy and Levelized Cost of Storage*, October 19, 2020. Accessed March 5, 2022, https://www.lazard.com/ perspective/levelized-cost-of-energy-and-levelized-cost-of-storage-2020/
13 Bakke, *The Grid*, 5.
14 "What a Ten-Year-Old Duck Can Teach Us About Electricity." *The Economist*, March 28, 2018.
15 Jenkins, Luke, and Thernstrom, "Getting to Zero Carbon Emissions in the Electric Power Sector," 2506–7.
16 NREL, *Renewable Electricity Futures Study, Executive Summary*, 2012. Accessed July 22, 2020, https://www.nrel.gov/docs/fy13osti/52409-ES. pdf
17 Ibid., iii.
18 Ibid., 14.
19 United Nations Development Program, "The People's Climate Vote" (2021). Accessed December 22, 2021, https://www.undp.org/publications/ peoples-climate-vote
20 Peter Kelly-Detwiler, "Renewable Energy in Spain: The Good, and the Downright Ugly." *Forbes*, May 8, 2013. Accessed December 24, 2021, https://www.forbes.com/sites/peterdetwiler/2013/05/08/

renewable-energy-in-spain-the-good-and-the-downright-ugly/
?sh=31d21e453899

21 Jenkins, Luke, and Thernstrom, "Getting to Zero Carbon Emissions in the Electric Power Sector," 2508.

22 Ibid., 2509.

23 "Top District Heating Countries – Euroheat and Power 2015 Survey Analysis." Solar Thermal World, July 1, 2016. Accessed October 16, 2021, https://www.solarthermalworld.org/news/top-district-heating-countries-euroheat-power-2015-survey-analysis

24 See the IEA's Tracking Report on "Heat Pumps." November 2021. Accessed March 5, 2022, https://www.iea.org/reports/heat-pumps

25 Natalie Filatoff, "New US Study Finds Renewable Energy Storage Costs Need to Drop 90%." PV Magazine, August 12, 2019.

26 Sarah Marie Jordaan and Kavita Surana, "We Calculated Emissions Due to Electricity Loss on the Power Grid – Globally It's a Lot." The Conversation, December 11, 2019. Accessed June 30, 2021, https://theconversation.com/we-calculated-emissions-due-to-electricity-loss-on-the-power-grid-globally-its-a-lot-128296

27 Center for Climate and Energy Solutions, Microgrid Momentum: Building Efficient, Resilient Power (March 2017). Microgrids are "relatively small, controllable power systems comprised of one or more generation units connected to nearby users that can be operated with, or independently from, the local bulk (i.e. high-voltage) transmission system."

28 An overview is ClearPath, "Introduction to Energy Storage." Accessed May 20, 2021, https://clearpath.org/tech-101/intro-to-energy-storage/

29 IRENA, Electricity Storage and Renewables: Costs and Markets to 2030 (2017), 4. Accessed April 3, 2022, https://www.irena.org/publications/2017/Oct/Electricity-storage-and-renewables-costs-and-markets

30 Micah S. Ziegler et al., "Storage Requirements and the Cost of Reshaping Renewable Energy Toward Grid Decarbonization." Joule 3 (9), 2134–53.

31 Ibid., 2134.

32 Dynamic pricing sends signals enabling customers to "decide whether to continue consumption at higher prices or to cut electricity usage during peak times." Marilyn Brown and Shan Zhou, "Smart Grid Policies: An International Review." Georgia Tech School of Public Policy, Working Paper 70 (2012), 11.

33 Moslem Uddin et al., "A Review on Peak Load Shaving Strategies." Renewable and Sustainable Energy Reviews 82 (Part 3) (2018), 3323–32, here 3323.

34 From smartgrid.gov. Accessed June 22, 2021, https://www.smartgrid.gov/the_smart_grid/smart_grid.html

35 Ibid.

36 Heinberg and Fridley, *Our Renewable Future*, 60.

37 Johann J. Kranz and Arnold Picot, "Toward an End-to-End Smart Grid: Overcoming Bottlenecks to Facilitate Competition and Innovation in Smart Grids." National Regulatory Research Institute, 2011. Accessed July 1, 2021, https://www.uni-goettingen.de/de/document/download/ 6e55d30e5d95c32d83088f34c6dcac42.pdf/Towards%20an%20 End-to-End%20Smart%20Grid%20Overcoming%20Bottlenecks%20 to%20Facilitate%20Competition%20and%20Innovation%20in%20 Smart%20Grids_Kranz%20Picot_2011.pdff

38 Environmental Defense Fund, *Grid Modernization: The Foundation for Climate Change Progress* (2017). Accessed 3 April, 2022, https://www. edf.org/sites/default/files/GridModReport.pdf

39 EPRI, *Estimating the Costs and Benefits of the Smart Grid* (2011), 1–4 to 1–8. This report is available from the US Department of Energy Smart Grid site. Accessed March 30,2022, https://www.smartgrid.gov/ files/documents/Estimating_Costs_Benefits_Smart_Grid_Preliminary_ Estimate_In_201103.pdf

40 Malcolm Abbot, "Is the Security of Electricity Supply a Public Good?" *The Electricity Journal* (August/September 2001), 33.

41 For a review of drivers and barriers, see Brown and Zhou, "Smart Grid Policies: An International Review."

42 See Scott Victor Valentine, Marilyn A. Brown, and Benjamin K. Sovacool, *Empowering the Great Energy Transition: Policy for a Low-Carbon Future* (Columbia University Press, 2019), 93–6.

43 Luis Camarinha-Matos, "Collaborative Smart Grids: A Survey in Trends." *Renewable and Sustainable Energy Reviews* 65 (2016), 283.

44 Ibid., 288.

45 This EV overview is based on the IEA's *Global EV Outlook: Towards Trans-Modal Electrification* (2018), particularly the Executive Summary. Accessed June 14, 2020, https://www.iea.org/reports/ world-energy-outlook-2018

46 Felix Richter, "Which Countries Have the Most Electric Cars?" World Economic Forum, February 13, 2021. Accessed June 30, 2021, https://www.weforum.org/agenda/2021/02/electric-vehicles-europe- percentage-sales/

47 From the Overview in the IEA's *Global EV Outlook 2021* (April 2021). Accessed July 1, 2021, https://www.iea.org/reports/global-ev-outlook- 2021?mode=overview

48 Sandra Wappelhorst, *The End of the Road? An Overview of Combustion-Engine Car Phase-Out Announcements Across Europe*, International Council on Clean Transportation (May 2020).

49 Charles Riley, "The Great Electric Car Race Is Just Beginning." *CNN Business*, August 2019, https://www.cnn.com/interactive/2019/08/ business/electric-cars-audi-volkswagen-tesla/?mod=article_inline

50 "Here is Newsom's Plan to Allow Only Zero-Emissions Car Sales in

California By 2035." *Los Angeles Times*, September 23, 2020. Medium and heavy-duty trucks are banned as of 2045.

51 Overview, IEA's *Global EV Outlook 2021*.

52 Heinberg and Fridley, *Our Renewable Future*, 24.

53 BNEF, "Battery Pack Prices Cited Below $100/kWh for the First Time in 2020, While Market Average Sits at $137/kWh," December 16, 2020. Accessed March 31, 2022, https://about.bnef.com/blog/battery-pack-prices-cited-below-100-kwh-for-the-first-time-in-2020-while-market-average-sits-at-137-kwh/

54 Jasper Jolly, "Electric Cars as Cheap to Manufacture as Regular Models by 2024." *The Guardian*, October 21, 2020.

55 Mark Kane, "Global Plug-In Car Sales September 2021: Doubled to New Record." *Inside EVs*, November 1, 2021. Accessed December 24, 2021, https://insideevs.com/news/544743/global-plugin-car-sales-september2021/

56 Enrique Dans, "For Elon Musk, Economies of Scale Are Not Rocket Science … Or Are They?" *Forbes*, June 5, 2020, https://www.forbes.com/sites/enriquedans/2020/06/05/for-elon-musk-economies-of-scale-are-not-rocket-science-or-arethey/?sh=5aef34a75316

57 Ibid.

58 Economist Intelligence Unit, *Democracy Index 2020: In Sickness and In Health*, 2020. Accessed January 5, 2021, https://www.eiu.com/n/campaigns/democracy-index-2020/

59 Henry Sanderson, "Congo, Child Labour, and Your Electric Car." *Financial Times*, July 7, 2019.

60 Kang Miao Tau et al., "Integration of Electric Vehicles in Smart Grid: A Review on Vehicle to Grid Technologies and Optimization Techniques." *Renewable and Sustainable Energy Reviews* 53 (2016), 720–32.

61 David Reichmuth, "Are Electric Vehicles Really Better for the Climate? Yes. Here's Why." Union of Concerned Scientists. February 11, 2020. Accessed March 6, 2022, https://blog.ucsusa.org/dave-reichmuth/are-electric-vehicles-really-better-for-the-climate-yes-heres-why

62 A University of Toronto study reported by Russell Gold, Jessica Kuronen, and Elbert Wang, "Are Electric Cars Really Better for the Environment?" *Wall Street Journal*, March 22, 2021. Accessed June 15, 2021, https://www.wsj.com/graphics/are-electric-cars-really-better-for-the-environment/

63 Reichmuth, "Are Electric Vehicles Really Better for the Climate?"

64 This discussion draws on an overview by Samantha Gross, "The Challenge of Decarbonizing Heavy Transport." Brookings Institution, October 2020. Accessed April 3, 2022, https://www.brookings.edu/wp-content/uploads/2020/09/FP_20201001_challenge_of_decarbonizing_heavy_transport.pdf

65 Ibid., 5.

66 Heinberg and Fridley, *Our Renewable Future*, 84.

67 The International Council on Clean Transportation, *Long-Term Aviation Fuel Decarbonization: Progress, Roadblocks, and Opportunities* (January 2019), especially pp. 1–6. The quote is from p. 3. Accessed April 3, 2022, https://theicct.org/publication/long-term-aviation-fuel-decarbonization-progress-roadblocks-and-policy-opportunities/

68 International Transport Forum, *Decarbonizing Marine Transport: Pathways to Zero-Carbon Shipping By 2035* (OECD, 2018), 7. Accessed April 3, 2022, https://www.itf-oecd.org/sites/default/files/docs/decarbonising-maritime-transport-2035.pdf

69 Katherine Hamilton and Tammy Ma, "Electric Aviation Could Be Closer Than You Think." *Scientific American,* November 10, 2020. Accessed May 5, 2021, https://www.scientificamerican.com/article/electric-aviation-could-be-closer-than-you-think/

70 McKinsey, *Decarbonization of Industrial Sectors: The Next Frontier,* June 2018, 6. Accessed March 6, 2022, https://www.mckinsey.com/~/media/mckinsey/business functions/sustainability/our insights/how industry can move toward a low carbon future/decarbonization-of-industrial-sectors-the-next-frontier.pdf. Feedstocks are "the raw materials that companies process into individual products."

71 Ibid.

72 Jeffrey Rissman et al. "Technologies and Polices to Decarbonize Global Industry: Review and Assessment of Mitigation Drivers Through 2070." *Applied Energy* 266 (May 2020), 114848.

73 UN Department of Economic and Social Affairs, Sustainable Development, SDG7. Accessed July 22, 2021, https://sdgs.un.org/goals/goal7

74 World Bank (with IEA and others), *Tracking SDG7: The Energy Progress Report 2020*. Accessed March 6, 2022, https://openknowledge.worldbank.org/handle/10986/33822

75 IEA, "Access to Electricity." Accessed December 34, 2021, https://www.iea.org/reports/sdg7-data-and-projections/access-to-electricity

76 Sadie Cox et al., *Distributed Generation to Support Development-Focused Climate Action*, NREL and the US Agency for International Development (AID), September 2016, 2.

77 India Smart Grid Knowledge Portal. Accessed May 5, 2021, https://indiasmartgrid.org/Distributed-Generation.php

7 Hard Choices and an Opportunity

1 Robin Cowan, "Nuclear Power Reactors: A Study in Technological Lock-In." *The Journal of Economic History* 50 (3) 1990, 541–67.

2 Ibid.

3 Sivaram, *Taming the Sun*, 21–3.

4 Although there is a movement of "Green for Nuclear Energy" that sees it

as essential given catastrophic risks of climate change. See https://www.greensfornuclear.energy. Accessed July 4, 2021.

5 Steve Clemmer, Jeremy Richardson, Sandra Sattler, and Dave Lockbaum, "The Nuclear Power Dilemma: Declining Profits, Plant Closures, and the Threat of Rising Carbon Emissions." Union of Concerned Scientists, November 2018. Accessed May 22, 2020, https://www.ucsusa.org/sites/default/files/attach/2018/11/Nuclear-Power-Dilemma-full-report.pdf

6 Akshat Rathi, "Europe's Heatwave Is Forcing Nuclear Power Plants to Shut Down." *Quartz*, August 6, 2018.

7 Matthew Bandyk, "For Nuclear Plants Operating on Thin Margins, Growing Climate Risks Prompt Tough Choices." *Utility Dive*, September 10, 2020. For a detailed discussion, see Sarah M. Jordan, Afreen Siddiqui, William Kakenmaster, and Alice M. Hill, "The Climate Vulnerabilities of Nuclear Power." *Global Environmental Politics* 19 (4) 2019, 3–13.

8 This draws on the World Nuclear Association's "Nuclear Power in the World Today." Updated March 2020. Accessed June 2020, https://www.world-nuclear.org/information-library/current-and-future-generation/nuclear-power-in-the-world-today.aspx?sms_ss=blogger&at_xt=4da1a75c7d3c5d10%252C0

9 Ibid. This also draws on the World Nuclear Association's "Nuclear Power in the USA." Updated January 2021. Accessed May 2, 2021, https://www.world-nuclear.org/information-library/country-profiles/countries-t-z/usa-nuclear-power.aspx

10 Jochen Bittner, "The Tragedy of Germany's Energy Experiment." *New York Times*, January 8, 2020.

11 Daniel Oberhaus, "Germany Rejected Nuclear Power – and Deadly Emissions Spiked." *Wired*, January 23, 2020. Accessed March 6, 2022, https://www.wired.com/story/germany-rejected-nuclear-power-and-deadly-emissions-spiked/

12 The full study was Stephen Jarvis, Olivier Deschenes, and Akshaya Jha, "The Private and External Costs of Germany's Nuclear Phaseout." National Bureau of Economic Research, December 2019. Accessed March 6, 2022, https://www.nber.org/papers/w26598

13 IEA, *Nuclear Power in a Clean Energy System*, 2019. Accessed July 2, 2021, https://www.iea.org/reports/nuclear-power-in-a-clean-energy-system

14 "How Much Nuclear Power Is at Risk of Retirement?" *Carbon Brief*, July 24, 2018. Accessed March 6, 2022, https://www.carbonbrief.org/mapped-the-us-nuclear-power-plants-at-risk-of-shutting-down

15 From World Nuclear Association, "Nuclear Power in the USA."

16 Bittner, "The Tragedy of Germany's Energy Experiment."

17 See the Third Way's "Advanced Nuclear 101." Accessed March 6, 2022, https://www.thirdway.org/report/advanced-nuclear-101

18 World Nuclear Association, Advanced Nuclear Power Reactors, most recent update April 2021. Accessed April 3, 2022, https://world-nuclear.

org/information-library/nuclear-fuel-cycle/nuclear-power-reactors/
advanced-nuclear-power-reactors.aspx. See also NA, Small Nuclear
Power Reactors, updated December 2021. Accessed April 3, 2022,
https://world-nuclear.org/information-library/nuclear-fuel-cycle/nuclear-
power-reactors/small-nuclear-power-reactors.aspx

19 Ben Soltoff, "Advanced Nuclear: A Climate-Tech Comeback Story."
    Greenbiz, November 13, 2020. Accessed December 5, 2020, https://www.
    greenbiz.com/article/advanced-nuclear-climate-tech-comeback-story

20 Scott Carpenter, "Bill Gates' Nuclear Start-Up Reveals Mini-Reactor
    Design Including Molten Energy Storage." *Forbes*, August 31, 2020.

21 Soltoff, "Advanced Nuclear: A Climate-Tech Comeback Story."

22 Karin Backstrand, James Meadowcroft, and Michael Oppenheimer,
    "The Politics and Policy of Carbon Capture and Storage: Framing
    an Emerging Technology." *Global Environmental Change* 21 (2011),
    275–81.

23 From the IPCC's Special Report on *Carbon Dioxide Capture and
    Storage: Summary for Policy Makers* (Cambridge University Press,
    2005), 3.

24 Manuel S. Godov et al., "About How to Capture and Exploit the CO2
    Surplus That Nature, Per Se, Is Not Capable of Fixing." *Microbial
    Biotechnology* 10 (5) (2017), 1216–25.

25 IEA, *20 Years of Carbon Capture and Storage: Accelerating Future
    Deployment* (2016), 57. Accessed March 7, 2022, https://www.iea.org/
    reports/20-years-of-carbon-capture-and-storage

26 On parallels with other technologies, see Varun Rai, David G. Victor,
    and Mark C. Thurber, "Carbon Capture and Storage at Scale: Lessons
    from the Growth of Analogous Energy Technologies." *Energy Policy* 38
    (2010), 4089–98.

27 Technology-based regulation is described in Daniel J. Fiorino, *The New
    Environmental Regulation* (MIT Press, 2006).

28 David Vogel, *California Greenin': How the Golden State Became an
    Environmental Leader* (Princeton University Press, 2018).

29 Vincent Gonzales, Alan Krupnick, and Lauren Dunlap, "Carbon Capture
    and Storage 101." Resources for the Future, May 6, 2020. Accessed June
    10, 2020, https://www.rff.org/publications/explainers/carbon-capture-
    and-storage-101/

30 Clearpath, "Carbon Capture 101." Accessed May 6, 2020, https://
    clearpath.org/tech-101/carbon-capture-101/

31 Carbon Capture and Storage Association, "What Is CCS?" Accessed
    May 7, 2020, https://www.ccsassociation.org/

32 Sharon Kelly, "How America's Clean Coal Dream Unraveled." *The
    Guardian*, March 2, 2018.

33 Ibid. Also see Ian Urbina, "Piles of Dirty Secrets Behind a Model 'Clean
    Coal' Project." *New York Times*, July 5, 2016.

34 David Hawkins and George Peridas, "Kemper County IGCC: Death Knell

for Carbon Capture? NOT." Natural Resources Defense Council Expert Blog, July 28, 2017. Accessed March 7, 2022, https://www.nrdc.org/experts/george-peridas/kemper-county-igcc-death-knell-carbon-capture-not

35 Carlos Anchondo and Edward Klump, "Petra Nova Is Closed: What It Means for Carbon Capture." *E&E News*, September 22, 2020. Accessed July 5, 2021, https://www.eenews.net/articles/petra-nova-is-closed-what-it-means-for-carbon-capture/

36 Institute for Energy Economics and Financial Analysis, *Holy Grail of Carbon Capture and Storage Continues to Elude Coal Industry*, November 2018. Accessed July 20, 2020, https://ieefa.org/wp-content/uploads/2018/11/Holy-Grail-of-Carbon-Capture-Continues-to-Elude-Coal-Industry_November-2018.pdf

37 See Kelly Levin, James Mulligan, and Gretchen Ellison, "Taking Greenhouse Gases From the Sky: 7 Things to Know About Carbon Removal." World Resources Institute, March 2018. Accessed August 11, 2020, https://www.wri.org/blog/2018/03/taking-greenhouse-gases-sky-7-things-know-about-carbon-removal

38 IEA, *20 Years of Carbon Capture and Storage*, 10.

39 Ibid., 103.

40 An excellent resource on direct air capture and other carbon removal methods is American University's Institute for Carbon Removal Law and Policy, including a fact sheet on "What Is Carbon Removal?" Accessed July 21, 2021, https://www.american.edu/sis/centers/carbon-removal/fact-carbon-removal.cfm

41 Global CCS Institute, *Bioenergy and Carbon Capture and Storage*, 2019, 5. Accessed October 4, 2020, https://www.globalccsinstitute.com/wp-content/uploads/2019/03/BECCS-Perspective_FINAL_PDF.pdf

42 Ibid., 10.

43 Mathilde Fajardy and Niall McDowell, "The Energy Return on Investment of BECCS: Is BECCS a Threat to Energy Security?" *Energy and Environmental Science* 11 (2018), 1581–99.

44 Institute for Carbon Removal Law and Policy, "What Is Direct Air Capture?" Accessed July 21, 2021, https://www.american.edu/sis/centers/carbon-removal/fact-sheet-direct-air-capture.cfm

45 Sabine Fuss et al., "Betting on Negative Emissions." *Nature Climate Change* 4 (2014), 852. The original report is Climate Change 2014: Mitigation of Climate Change. Accessed April 2, 2022, https://www.ipcc.ch/report/ar5/wg3/

46 Abby Rabinowitz and Amanda Simson, "The Dirty Secret of the World's Plan to Avert Climate Disaster." *Wired*, December 18, 2017. Accessed September 15, 2020, https://www.wired.com/story/the-dirty-secret-of-the-worlds-plan-to-avert-climate-disaster/

47 US Department of Energy, Office of Energy Efficiency and Renewable Energy, "10 Things You Might Not Know About Hydrogen and Fuel Cells."

October 8, 2019. Accessed April 3, 2022, https://www.energy.gov/eere/articles/10-things-you-might-not-know-about-hydrogen-and-fuel-cells

48 Marc A. Rosen and Seema Koohi-Fayegh, "The Prospects for Hydrogen as an Energy Carrier: An Overview of Hydrogen Energy and Hydrogen Energy Systems." *Energy, Ecology, and Environment* 1 (2016), 10–29.

49 IRENA, *Green Hydrogen: A Guide to Policy Making*, 2020, 6. Accessed June 5, 2021, https://www.irena.org/-/media/Files/IRENA/Agency/Publication/2020/Nov/IRENA_Green_hydrogen_policy_2020.pdf

50 See Rosen and Koohi-Fayegh, "The Prospects for Hydrogen," and the IRENA reports on green hydrogen cited in this chapter.

51 IRENA, *Hydrogen From Renewable Power: Technology Outlook for the Energy Transition*, 2018, 38. Accessed April 3, 2022, https://www.irena.org/-/media/Files/IRENA/Agency/Publication/2018/Sep/IRENA_Hydrogen_from_renewable_power_2018.pdf

52 Ibid.

53 Ibid.

54 Giles Parkinson, "The Long-Term Energy Storage Challenge: Batteries Not Included." *Greentech Media*, December 17, 2013. Accessed March 9, 2022, https://www.greentechmedia.com/articles/read/the-long-term-storage-challenge-batteries-not-included

55 "New and Large Scale Hydrogen Hub to Support Denmark's Green Transition." *FuelCellsWorks*, December 1, 2020. Accessed March 9, 2022, https://fuelcellsworks.com/news/new-and-large-scale-hydrogen-hub-to-support-denmarks-green-transition/

56 Deloitte/Ballard, *Fueling the Future of Mobility: Hydrogen and Fuel Cell Solutions for Transportation*, 2020, 5. Accessed June 22, 2021, https://www.ballard.com/about-ballard/newsroom/news-releases/2020/01/08/deloitte-ballard-joint-white-paper-assesses-hydrogen-fuel-cell-solutions-for-transportation

57 Susan Phillips, "Japan Is Betting Big on the Future of Hydrogen Cars." *NPR*, March 18, 2019.

58 Jason Deign, "Hydrogen Mobility: Coming Soon to a Bus or Truck Near You?" *Greentech Media*, March 6, 2020. Accessed July 5, 2020, https://www.greentechmedia.com/articles/read/hydrogen-mobility-coming-soon-to-a-bus-or-truck-near-you

59 Deloitte/Ballard, *Fueling the Future of Mobility*, 16–18.

60 Ibid., 1–2.

61 Ibid., 1.

62 Rosen and Koohi-Fayegh, "The Prospects for Hydrogen," 15.

63 IRENA, *Hydrogen From Renewable Power*.

64 IRENA, *Green Hydrogen: A Guide to Policy Making*, 10–13.

65 IEA, *Nuclear Power in a Clean Energy System*, 1.

66 IEA, *Tracking Clean Energy Progress*. Accessed March 9, 2022, https://www.iea.org/topics/tracking-clean-energy-progress

8 Accelerating the Energy Transition

1 This draws on Erik Figenbaum, "Perspectives on Norway's Supercharged Electric Vehicle Program." *Environmental Innovation and Societal Transitions* 25 (2017), 14–38.

2 Richter, "Chart: Which Countries Have the Most Electric Cars?"

3 Figenbaum, "Perspectives on Norway's Supercharged Electric Vehicle Policy," 16.

4 Ibid., 31.

5 Adapted from Stephen Ansolabehere and David M. Konisky, *Cheap and Clean: How Americans Think About Energy in the Age of Global Warming* (MIT Press, 2014), 9–10.

6 For an introduction, see Michael E. Kraft and Scott R. Furlong, *Public Policy: Politics, Analysis, and Alternatives*, 7th ed. (CQ Press, 2020).

7 On policies at different stages, see OECD/IEA, *A Policy Strategy for Carbon Capture and Storage* (2012) and IRENA's *Green Hydrogen: A Guide to Policy Making* (2020).

8 Jim Tankersley, "Biden Details $2 Trillion Plan to Rebuild Infrastructure and Reshape the Economy." *New York Times*, April 15, 2021.

9 Kelly Simms Gallagher, John P. Holdren, and Ambuj D. Sagar, "Energy Technology Innovation." *Annual Review of Environment and Resources* 31 (2006), 193–237.

10 IRENA, *Global Energy Transformation: A Roadmap to 2050* (2019), 9. Accessed April 3, 2022, https://www.irena.org/publications/2019/Apr/Global-energy-transformation-A-roadmap-to-2050-2019Edition

11 Daniel J. Fiorino and Manjyot Ahluwalia, "Regulating by Performance, Not Prescription: The Use of Performance Standards in Environmental Policy." In David M. Konisky (ed.), *Handbook of U.S. Environmental Policy* (Edward Elgar, 2020), 217–30.

12 On subsidies, see Johannes Urpelainen and Elisha George, "Reforming Fossil Fuel Subsidies: How the United States Can Restart International Cooperation." Brookings Institution, July 14, 2021. Accessed July 22, 2021, https://www.brookings.edu/research/reforming-global-fossil-fuel-subsidies-how-the-united-states-can-restart-international-cooperation/

13 Richard Schmalensee and Robert N. Stavins, "Lessons Learned From Three Decades with Cap and Trade." *Review of Environmental Economics and Policy* 11 (2017), 59–79.

14 Marco Guerzoni and Emilio Raiteri, "Demand-side vs. Supply-side Technology Policies: Hidden Treatment and New Empirical Evidence on the Policy Mix." *Research Policy* 44 (2015), 726–47.

15 Hiroko Tabuchi, "Inside Conservative Groups' Efforts to 'Make Dishwashers Great Again,'" *New York Times*, September 17, 2019.

16 Kenneth Gillingham, Richard Newell, and Karen Palmer, "Energy

Efficiency Policies: A Retrospective Examination." *Annual Review of Environment and Resources* 31 (2006), 161–92.

17 Sadie Cox, *Building Energy Codes: Policy Overview and Good Practices* (NREL, 2016), 1. Accessed July 8, 2020, https://www.nrel.gov/docs/fy16osti/65542.pdf

18 Sanya Carley, Elizabeth Baldwin, Lauren M. McLean, and Jennifer N. Bass, "Global Expansion of Renewable Energy Generation: An Analysis of Policy Instruments." *Environmental and Resource Economics* 68 (2017), 397–440.

19 National Conference of State Legislatures, "State Renewable Portfolio Standards and Goals." April 7, 2021. Accessed October 7, 2021, https://www.ncsl.org/research/energy/renewable -portfolio-standards.aspx

20 See EIA, "Feed-In Tariff: A Policy Tool Encouraging Deployment of Renewable Energy Technologies." May 30, 2013. Accessed June 9, 2021, https://www.eia.gov/todayinenergy/detail.php?id=11471

21 Virginia McConnell and Benjamin Leard, "Pushing Technology Into the Market: California's Zero Emission Vehicle Mandate." *Review of Environmental Economics and Policy* 15 (2021), 169–79.

22 Sonja Carley, "State Renewable Energy Electricity Policies: An Empirical Evaluation." *Energy Policy* 37 (2009), 3071–81.

23 Michael Greenstone and Ishan Nath, "Do Renewable Portfolio Standards Deliver Cost-Effective Carbon Abatement?" Energy Policy Institute at the University of Chicago (Working Paper 2019–62, 2020). Accessed March 9, 2022, https://bfi.uchicago.edu/working-paper/do-renewable-portfolio-standards-deliver-cost-effective-carbon-abatement/

24 Zifei Yang, Peter Slowik, Nic Lutsey, and Stephanie Seoule, *Principles for Effective Electric Vehicle Design* (ICCT, 2016), 2. Accessed March 9, 2022, https://theicct.org/sites/default/files/publications/ICCT_IZEV-incentives-comp_201606.pdf

25 Ibid., 4. Also see Christiane Munzel et al., "How Large Is the Effect of Financial Incentives on Electric Vehicle Sales? A Global Review and European Analysis." *Energy Economics* 84 (2019), 1–21.

26 Barry G. Rabe, *Can We Price Carbon?* (MIT Press, 2018).

27 World Bank, *State and Trends of Carbon Pricing 2021* (May 2021). Accessed July 10, 2021, https://openknowledge.worldbank.org/bitstream/handle/10986/35620/9781464817281.pdf?sequence=12&isAllowed=y

28 Ibid.

29 Carbon Pricing Leadership Coalition, "Should Every Country on Earth Adopt Sweden's Carbon Tax?" Accessed March 9, 2022, https://www.carbonpricingleadership.org/blogs/2019/10/18/should-every-country-on-earth-copy-swedens-carbon-tax. For more detail, see Julius J. Andersson, "Carbon Taxes and CO2 Emissions: Sweden as a Case Study." *American Economic Journal: Economic Policy* 11 (2019), 1–30.

30 World Bank, *State and Trends of Carbon Pricing 2021*, 12.

31 Government of Canada, "Carbon Pollution Pricing Systems Across

Canada." July 12, 2021. Accessed July 22, 2021, https://www.canada. ca/en/environment-climate-change/services/climate-change/pricing-pollution-how-it-will-work.html

32 Brian Murray and Nicholas Rivers, "British Columbia's Revenue Neutral Carbon Tax: A Review of the Latest 'Grand Experiment' in Environmental Policy." *Energy Policy* 86 (2015), 674–83.

33 Joseph E. Aldy and Robert N. Stavins, "The Promise and Problems of Carbon Pricing: Theory and Experience." *Journal of Environment and Development* 21 (2012), 152–80.

34 Center for Climate and Energy Solutions, "California Cap and Trade." Accessed June 2, 2021, https://www.c2es.org/content/california-cap-and-trade/

35 *California Climate Investments Using Cap-and-Trade Auction Proceeds* (March 2020). Accessed January 4, 2021, http://ww2.arb.ca.gov/sites/default/files/auction-proceeds/2020_cci_annual_report.pdf

36 Sarah Kaplan, "Biden Wants an All-Electric Federal Fleet." *Washington Post*, January 28, 2021.

37 Future Policy, "Japan's Top Runner Programme." Accessed January 14, 2021, https://www.futurepolicy.org/climate-stability/japans-top-runner-programme/

38 European Commission, "About the Energy Label and Ecodesign." Accessed January 17, 2021, https://ec.europa.eu/info/energy-climate-change-environment/standards-tools-and-labels/products-labelling-rules-and-requirements/energy-label-and-ecodesign/about_en

39 California Energy Commission, "SB 100 Joint Agency Report." Accessed March 9, 2022, https://www.energy.ca.gov/sb100

40 California Environmental Protection Agency, "Climate Action Programs." Accessed October 22, 2021, https://calepa.ca.gov/climate/

41 Aklin and Urpelainen, *Renewables: The Politics of a Global Energy Transition*, 38–40.

42 Niels I. Meyer, "Learning from Wind Energy Policy in the EU: Lessons from Denmark, Sweden, and Spain." *Environmental Policy and Governance* 17 (2007), 351.

43 Ibid., 349.

44 Heinberg and Fridley, *Our Renewable Future*, 75.

45 Andreas Cala, "Renewable Energy in Spain Is Taking a Beating." *New York Times*, October 8, 2013.

46 Ibid.

47 Patricia Martinez Alonso et al., "Losing the Roadmap: Renewable Energy Paralysis in Spain and Its Implications for the EU Low Carbon Economy." *Renewable Energy* 89 (2016), 681.

48 Jason Deign, "Spain Is a Case Study in How Not to Foster Renewables." *Greentech Media*, May 5, 2017. Accessed June 17, 2020, https://www.greentechmedia.com/articles/read/spain-is-a-case-study-in-how-not-to-foster-renewables

49 IEA, *Africa Energy Outlook 2019. Overview: Kenya*. Accessed July 16, 2021, https://iea.blob.core.windows.net/assets/44389eb7–6060–4640–91f8–583994972026/AEO2019_KENYA.pdf

50 IEA, "SDG7 Data and Projections: Access to Electricity." Accessed July 17, 2021, https://www.iea.org/reports/sdg7-data-and-projections/access-to-electricity

51 Johnny Wood, "Kenya Is Aiming to Be Powered Entirely by Green Energy by 2020." World Economic Forum, December 5, 2018. Accessed July 16, 2021, https://www.weforum.org/agenda/2018/12/kenya-wants-to-run-entirely-on-green-energy-by-2020/

52 Roy Janho, "Renewable Energy in Kenya: An Examination of the Legal Instruments and Institutional Changes That Successfully Attracted Foreign Investment." *Energy Central*, November 18, 2020. Accessed July 14, 2021, https://energycentral.com/c/pip/renewable-energy-kenya-examination-legal-instruments-and-institutional-changes

53 William F. Lamb and Jan C. Minx, "The Political Economy of Climate Policy: Architectures of Constraint and Typology of Countries." *Energy Resources and Social Science* 64 (2020), 2.

54 Daniela Stevens, "The Influence of the Fossil Fuel and Emission-Intensive Industries on the Stringency of Mitigation Policies: Evidence from the OECD Countries and Brazil, Russia, India, Indonesia, China, and South Africa." *Environmental Policy and Governance* 29 (2019), 279–92.

55 R. Kent Weaver and Bert Rockman, eds., *Do Institutions Matter? Government Capabilities in the United States and Abroad* (Brookings Institution, 1993).

56 Lamb and Minx, "The Political Economy of Climate Policy," 7.

57 Duncan Liefferink et al., "Leaders and Laggards in Environmental Performance: A Quantitative Analysis of Domestic Policy Outputs." *Journal of European Public Policy* 16 (2009), 677–700.

58 Elizabeth Baldwin, Sanya Carley, and Sean Nicholson-Crotty, "Why Do Countries Emulate Each Other's Policies? A Global Study of Renewable Energy Diffusion." *World Development* 120 (2019), 30.

59 Xavier Labandeira, Jose M. Labeaga, Pedro Linares, and Xiral Lopez-Otero, "The Impacts of Energy Efficiency Policies: Meta-Analysis." *Energy Policy* 147 (December 2020), 8. They found efficiency policies cut demand by, on average, 8–10% in all studies and 3–5% in studies they rated as highly robust.

60 Rohan Best, Paul J. Burke, and Frank Jotzo, "Adoption of Wind and Solar Energy: The Roles of Carbon Pricing and Adequate Policy Support." *Energy Policy* 118 (2018), 404–17.

61 Rohan Best, Paul J. Burke, and Frank Jotzo, "Carbon Pricing and Efficacy: Cross-Country Evidence." *Environmental and Resource Economics* 77 (2020), 69–94.

62 Best et al., "Adoption of Wind and Solar Energy," 404.

63 Kim Schumacher and Zhuoxiang Yang, "The Determinants of Wind Energy Growth in the United States: Drivers and Barriers to State-Level Development," *Renewable and Sustainable Energy Reviews* 97 (2018), 1–13.

64 On the effects of different policies, see Nick Johnstone, Ivan Hascic, and David Popp, "Renewable Energy Policies and Technical Innovation: Evidence Based on Patent Counts." *Environmental and Resource Economics* 45 (2010), 133–55.

65 Thomas Dietz, Christina Leshko, and Aaron M. McCright, "Politics Shapes Individual Choices About Energy Efficiency." *PNAS* 110 (2013), 9191–2.

66 Hal Bernton, "Washington State's Carbon Pricing Bill Could Be Most Far-Reaching in Nation." *Seattle Times*, May 1, 2021.

67 Soren Anderson, Ioana Elena Marinescu, and Boris Shor, "Can Pigou at the Polls Stop Us Melting the Poles?" July 31, 2019. Accessed June 14, 2021, https://www.nber.org/system/files/working_papers/w26146/w26146.pdf

68 Nadia Popovich, "Climate Change Rises as a Public Priority. But It's More Partisan Than Ever." *New York Times*, February 20, 2020.

69 Sanya Carley and Tyler R. Browne, "Innovative US Energy Policy: A Review of States' Policy Experience." *WIRE's Energy Environment* 2 (2013), 495.

70 Nick Johnstone, Ivan Hascic, and Maria Kalamova, "Environmental Policy Design Characteristics and Technological Innovation." *Economia Politica* 2 (2010), 277–302.

71 Jiaqi Liang and Daniel J. Fiorino, "The Implications of Policy Stability for Renewable Energy in the United States, 1974–2009." *Policy Studies Journal* 41 (1), 97–118.

72 Xavier Fernandez-I-Marin, Christoph Knill, and Yves Steinbach, "Studying Policy Design in Comparative Perspective." *American Political Science Review* 115 (3) 2021, 937.

73 Sonja Carley and Michelle Graff, "A Just U.S. Energy Transition." In David M. Konisky (ed.), *Handbook of U.S. Environmental Policy* (Edward Elgar, 2020), 434–47.

9 The Clean Energy Future

1 William B. Bonvillian and Charles Weiss, "Stimulating Innovation in Energy Technology." *Issues in Science and Technology* 26 (2009).

2 IEA, *Energy Technology Perspectives 2020: Accelerating Technology Progress for a Sustainable Future* (2020), 18. Accessed April 3, 2022, https://iea.blob.core.windows.net/assets/7f8aed40-89af-4348-be19-c8a67df0b9ea/Energy_Technology_Perspectives_2020_PDF.pdf

3 Jason Deign, "Is Wave Energy Ready to Climb Out of the "Valley of Death'?" *Greentech Media*, October 2, 2020. Accessed June 13, 2021,

https://www.greentechmedia.com/articles/read/is-wave-energy-ready-to-climb-out-of-the-valley-of-death-for-new-technologies

4 Robert Fri, "From Energy Wish Lists to Technological Realities." *Issues in Science and Technology* 23 (2006), 63–8.

5 Solyndra involved a government loan of $500 million to a PV firm that failed. See Federico Caprotti, "Protecting Innovative Niches in the Green Economy: The Rise and Fall of Solyndra, 2005–2011." *Geojournal* 82 (2017), 937–55.

6 Kelly Sims Gallagher et al. "The Energy Technology Innovation System." *Annual Review of Environment and Resources* 37 (2012), 137–62.

7 IEA, *Energy Technology Perspectives 2020: Special Report on Clean Energy Innovation* (2020). Accessed July 15, 2021, https://iea.blob.core.windows.net/assets/04dc5d08-4e45-447d-a0c1-d76b5ac43987/Energy_Technology_Perspectives_2020_-_Special_Report_on_Clean_Energy_Innovation.pdf

8 IEA, *Tracking Clean Energy Progress*, 2020. Accessed March 6, 2022, https://www.iea.org/topics/tracking-clean-energy-progress

9 Sivaram, *Taming the Sun*, 150–77.

10 Varun Sivaram, John O. Dabiri, and David M. Hart, "The Need for Continued Innovation in Solar, Wind, and Energy Storage." *Joule* 9 (2018), 1639–47.

11 Carley and Graff, "A Just U.S. Energy Transition," 436.

12 Jeffrey D. Sachs, *The Age of Sustainable Development* (Columbia University Press, 2015), 227.

13 Ariel Drehobl and Lauren Ross, *Lifting the High Energy Burden in America's Largest Cities*, ACEEE, 2016. Accessed May 15, 2021, https://www.aceee.org/research-report/u1602

14 Adam Chandler, "Where the Poor Spend More Than 10 Percent of Their Income on Energy." *The Atlantic*, June 8, 2018.

15 Sammy Roth, "California's Clean Energy Programs Are Mainly Benefiting the Rich, Study Finds." *Los Angeles Times*, June 25, 2020.

16 On coal areas, see Adam Mayer, "A Just Transition for Coal Miners? Community Identity and Support from Local Policy Actors." *Environmental Innovation and Societal Transformations* 28 (2018), 1–13.

17 Melissa Eddy, "Why Green Germany Remains Addicted to Coal." *New York Times*, October 10, 2019.

18 Johannes Urpelainen, "Fossil Fuels: Save the Workers, Kill the Industry." *The Hill*, April 30, 2020. Accessed August 15, 2021, https://thehill.com/opinion/energy-environment/494427-fossil-fuels-save-the-workers-kill-the-industry

19 Eddy, "Why Green Germany Remains Addicted to Coal."

20 European Commission, "The Just Transition Mechanism: Making Sure No One Is Left Behind." Accessed March 9, 2022, https://ec.europa.eu/info/strategy/priorities-2019–2024/european-green-deal/actions-being-taken-eu/just-transition-mechanism_en

21 Ibid.
22 James Marshall, "Duckworth Introduces Plan for Coal Country." *E&E Daily*, July 24, 2020. Accessed August 11, 2020, https://www.eenews. net/stories/1063618057
23 International Labor Organization, *World Employment and Social Outlook 2018: Greening With Jobs* (2018), 1 and 44. Accessed June 25, 2021, https://www.ilo.org/weso-greening/documents/WESO_Greening_ EN_web2.pdf
24 Conrad Kunze and Soren Becker, *Energy Democracy in Europe: A Survey and Outlook* (Rosa Luxembourg Stiftung, 2014).
25 Matthew J. Burke and Jennie C. Stephens, "Political Power and Renewable Energy Futures: A Critical Review." *Energy Research and Social Science* 35 (2018), 79.
26 Shannon Hall, "Exxon Knew About Climate Change Almost 40 Years Ago." *Scientific American*, October 26, 2015. Accessed July 12, 2021, https://www.scientificamerican.com/article/exxon-knew-about-climate-change-almost-40-years-ago/
27 Susana Soeiro and Maria Ferreira Dias, "Energy Cooperatives in Southern European Countries: Are They Relevant for Sustainability Targets?" *Energy Reports* 6 (supplement 1) (2020), 448–53.
28 Ben Martin, "Communal Ownership Drives Denmark's Wind Revolution." Green Economy Coalition, September 20, 2017. Accessed October 27, 2021, https://www.greeneconomycoalition.org/news-and-resources/people-power-denmarks-energy-cooperatives
29 Burke and Stephens, "Political Power and Renewable Energy Futures," 80 and 79.
30 Matthew J. Burke and Jennie C. Stephens, "Energy Democracy: Goals and Policy Instruments for Sociotechnical Transitions." *Energy Research and Social Science* 33 (2017), 35–48.
31 Stella Schaller and Alexander Carius, *Convenient Truths: Mapping Climate Agendas of Right-Wing Populist Parties in Europe*. Adelphi, 2019. Accessed March 9, 2022, https://www.adelphi.de/en/publication/ convenient-truths
32 Shelley Welton and Joel Eisen, "Clean Energy Justice: Charting an Emerging Agenda." *Harvard Environmental Law Review* 43 (2019), 321.
33 Gregory C. Unruh, "Understanding Carbon Lock-In." *Energy Policy* 28 (2000): 817–36.
34 IRENA, "Investment Needs for the Global Energy Transformation." Accessed December 22, 2021, https://www.irena.org/financeinvestment/ Investment-Needs
35 This is Aidan Farrow, Kathryn A. Miller, and Lauri Myllyvirta's estimate in *Toxic Air: The Price of Fossil Fuels* (Greenpeace, 2020).
36 Sarah Pralle, "Issue Framing and Agenda Setting." In David M. Konisky (ed.), *Handbook of U.S. Environmental Policy* (Edward Elgar, 2020), 108–25.

37 Daniel J. Fiorino, "Creating the Green Economy: Government, Business, and a Sustainable Future." In Norman J. Vig, Michael E. Kraft, and Barry G. Rabe (eds), *Environmental Policy: New Directions for the Twenty-First Century*, 11th ed. (CA: Sage, 2022), 323–44.

38 HBSC Global Research, *A Climate for Recovery: The Colour of Stimulus Goes Green*, 2009. Accessed March 9, 2022, https://www.globaldashboard.org/wp-content/uploads/2009/HSBC_Green_New_Deal.pdf. See also Edward B. Barbier, *A Global Green Deal: Rethinking the Economic Recovery* (Cambridge University Press, 2010).

39 An example is Devashree Saha and Joel Jaeger, *American's New Climate Economy: A Comprehensive Guide to the Economic Benefits of Climate Policy in the United States*, World Resources Institute, July 2020. Accessed March 9, 2022, https://files.wri.org/d8/s3fs-public/americas-new-climate-economy.pdf

40 European Commission, *The European Green Deal*. Accessed July 20, 2021, https://ec.europa.eu/info/strategy/priorities-2019–2024/european-green-deal_en

41 Jim Tankersley, "Biden Details $2 Trillion Plan to Rebuild Infrastructure and Reshape the Economy." *New York Times*, March 31, 2021.

42 Kate Larson et al., *Green Stimulus Spending in the World's Major Economies*, Rhodium Group, February 4, 2021, 2. Accessed March 22, 2021, https://rhg.com/wp-content/uploads/2021/02/2020-Green-Stimulus-Spending-in-the-Worlds-Major-Economies.pdf

43 Ibid.

44 Joel Jaeger, "Lessons From the Great Recession for COVID-19 Green Recovery." World Resources Institute, November 24, 2020. Accessed December 5, 2020, https://www.wri.org/insights/lessons-great-recession-covid-19-green-recovery

45 Lisa Friedman, "What Is the Green New Deal? A Climate Proposal, Explained." *New York Times*, February 21, 2019.

46 Lachlan Carey, "How to Think About the Green New Deal After Its First Piece of Legislation." Center for Strategic and International Studies, December 21, 2019. Accessed June 11, 2021, https://www.csis.org/analysis/how-think-about-green-new-deal-after-its-first-piece-legislation

47 Aklin and Urpelainen, *Renewables: The Politics of a Global Energy Transition*, 13.

48 Andrew Cheon and Johannes Urpelainen, "How Do Competing Interest Groups Influence Environmental Policy? The Case of Renewable Electricity in Industrialized Democracies, 1987–2007." *Political Studies* 61 (2013), 894.

49 "The First Big Energy Shock of the Green Era." *The Economist*, October 16, 2021.

50 Cornelia Fraune and Michelle Knodt, "Sustainable Energy Transformations in the Age of Populism, Post-Truth Politics, and Local Resistance." *Energy Research and Social Science* 43 (2018), 1–7.

Index

Page numbers in *italics* denote a table/figure